A User-Friendly Guide to
Multivariate Calibration and Classification

A User-Friendly Guide to
# Multivariate Calibration and Classification

Tormod Næs
*Matforsk, Ås, Norway, and
Institute of Mathematics, University of Oslo, Norway*

Tomas Isaksson
*Agricultural University of Norway, Ås, Norway*

Tom Fearn
*University College London, London, UK*

Tony Davies
*Norwich Near Infrared Consultancy, Norwich, UK*

NIR Publications
Chichester, UK

**NIR** *Publications*

Published by:
NIR Publications, 6 Charlton Mill, Charlton, Chichester, West Sussex
PO18 0HY, UK. Tel: +44-1243-811334, Fax: +44-1243-811711,
E-mail: info@nirpublications.com, Web: www.nirpublications.com

ISBN: 0 9528666 2 5

British Library Cataloguing-in-Publication Data
A catalogue record for this book is available from the British Library

© NIR Publications 2002

All rights reserved. Apart from any fair dealing for the purposes of research or private study, or criticism or review, as permitted under the UK's Copyright, Designs and Patents Act 1988, no part of this publication may be reproduced, stored or transmitted, in any form or by any means, except with the prior permission in writing of the publishers, or in the case of reprographic re-production, in accordance with the terms of licences issued by the Copyright Licensing Agency. Enquiries concerning reprduction outside these terms should be sent to the publishers at the address above.

Typeset in 12/14 pt Times using Corel VENTURA 8
Printed and bound in the UK by The Cromwell Press Ltd, Trowbridge, Wiltshire.

# Contents

1 **Preface** . . . . . . . . . . . . . . . . . . . . . 1
   1.1 Marginal notes . . . . . . . . . . . . . . . . . 3
   1.2 Acknowledgements . . . . . . . . . . . . . . . 3

2 **Introduction** . . . . . . . . . . . . . . . . . . 5
   2.1 Basic problems and their solutions . . . . . . . . 6

3 **Univariate calibration and the need for multivariate methods** . . . . . . . . . . . . . 11
   3.1 Classical vs inverse calibration . . . . . . . . . 11
   3.2 Why are multivariate methods needed? . . . . . 14

4 **Multicollinearity and the need for data compression** . . . . . . . . . . . . . . . . . . 19
   4.1 Multicollinearity . . . . . . . . . . . . . . . . . 19
   4.2 Data compression . . . . . . . . . . . . . . . . 22
   4.3 Overfitting versus underfitting . . . . . . . . . . 24

5 **Data compression by PCR and PLS** . . . . . . . 27
   5.1 General model structure for PCR and PLS . . . . 27
   5.2 Standardisation of $x$-variables . . . . . . . . . 29
   5.3 Principal component regression (PCR) . . . . . . 30
   5.4 Partial least squares (PLS) regression . . . . . . . 33
   5.5 Comparison of PCR and PLS regression . . . . . 35
   5.6 Continuum regression . . . . . . . . . . . . . . 37

6 **Interpreting PCR and PLS solutions** . . . . . . 39
   6.1 Loadings and scores in principal component analysis (PCA) . . . . . . . . . . . . . . . . . . 39
   6.2 Loadings and loading weights in PLS . . . . . . 42
   6.3 What kind of effects do the components correspond to? . . . . . . . . . . . . . . . . . . 43
   6.4 Interpreting the results from a NIR calibration . . 46

7 **Data compression by variable selection** . . . . 55
   7.1 Variable selection for multiple regression . . . . 55

| | | | |
|---|---|---|---|
| | 7.2 | Which criterion should be used for comparing models? | 56 |
| | 7.3 | Search strategies | 60 |
| | 7.4 | Using the jack-knife to select variables in PLS regression | 67 |
| | 7.5 | Some general comments | 69 |

## 8 Data compression by Fourier analysis and wavelets . . . 71
8.1 Compressing multivariate data using basis functions. . . . 71
8.2 The Fourier transformation. . . . 74
8.3 The wavelet transform . . . 84

## 9 Non-linearity problems in calibration . . . . . 93
9.1 Different types of non-linearities exist . . . . . 93
9.2 Detecting multivariate non-linear relations. . . . 95
9.3 An overview of different strategies for handling non-linearity problems . . . 97

## 10 Scatter correction of spectroscopic data . . . 105
10.1 What is light scatter? . . . 105
10.2 Derivatives . . . 107
10.3 Multiplicative scatter correction (MSC) . . . 114
10.4 Piecewise multiplicative scatter correction (PMSC) . . . 119
10.5 Path length correction method (PLC-MC) . . . 120
10.6 Orthogonal signal correction (OSC) . . . 122
10.7 Optimised scaling (OS) . . . 123
10.8 Standard normal variate method (SNV) . . . 124

## 11 The idea behind an algorithm for locally weighted regression . . . 127
11.1 The LWR method . . . 127
11.2 Determining the number of components and the number of local samples . . . 130
11.3 Distance measures . . . 131
11.4 Weight functions . . . 133
11.5 Practical experience with LWR . . . 135

## 12 Other methods used to solve non-linearity problems . . . . . . . . . . . . . . . . . . . . 137
12.1 Adjusting for non-linearities using polynomial functions of principal components . . . . . . . 137
12.2 Splitting of calibration data into linear subgroups . . . . . . . . . . . . . . . . . . . 140
12.3 Neural nets . . . . . . . . . . . . . . . . . 146

## 13 Validation . . . . . . . . . . . . . . . . . . . . . 155
13.1 Root mean square error . . . . . . . . . . . . 155
13.2 Validation based on the calibration set . . . . . 156
13.3 Prediction testing . . . . . . . . . . . . . . . 157
13.4 Cross-validation . . . . . . . . . . . . . . . . 160
13.5 Bootstrapping used for validation . . . . . . . 162
13.6 *SEP*, *RMSEP*, *BIAS* and *RAP* . . . . . . . . . 163
13.7 Comparing estimates of prediction error . . . . 166
13.8 The relation between *SEP* and confidence intervals . . . . . . . . . . . . . . . . . . . . 170
13.9 How accurate can you get? . . . . . . . . . . 172

## 14 Outlier detection . . . . . . . . . . . . . . . . 177
14.1 Why outliers? . . . . . . . . . . . . . . . . . 177
14.2 How can outliers be detected? . . . . . . . . . 178
14.3 What should be done with outliers? . . . . . . 189

## 15 Selection of samples for calibration . . . . . 191
15.1 Some different practical situations . . . . . . . 192
15.2 General principles . . . . . . . . . . . . . . . 193
15.3 Selection of samples for calibration using x-data for a larger set of samples . . . . . . . . 195

## 16 Monitoring calibration equations . . . . . . 201

## 17 Standardisation of instruments . . . . . . . . 207
17.1 General aspects . . . . . . . . . . . . . . . . 207
17.2 Different methods for spectral transfer . . . . . 209
17.3 Making the calibration robust . . . . . . . . . 216
17.4 Calibration transfer by correcting for bias and slope . . . . . . . . . . . . . . . . . . . . . . 217

## 18 Qualitative analysis/classification . . . . . . . 221

|      |                                                      |     |
|------|------------------------------------------------------|-----|
| 18.1 | Supervised and unsupervised classification           | 221 |
| 18.2 | Introduction to discriminant analysis                | 222 |
| 18.3 | Classification based on Bayes' rule                  | 225 |
| 18.4 | Fisher's linear discriminant analysis                | 230 |
| 18.5 | The multicollinearity problem in classification      | 234 |
| 18.6 | Alternative methods                                  | 243 |
| 18.7 | Validation of classification rules                   | 247 |
| 18.8 | Outliers                                             | 248 |
| 18.9 | Cluster analysis                                     | 249 |
| 18.10| The collinearity problem in clustering               | 259 |

## 19 Abbreviations and symbols . . . . . . . . . 261
- 19.1 Abbreviations . . . . . . . . . . . . . . . . . 261
- 19.2 Important symbols . . . . . . . . . . . . . . 262

## 20 References . . . . . . . . . . . . . . . . . . 265

## A Appendix A. Technical details . . . . . . . . 285
- A.1 An introduction to vectors and matrices . . . . 285
- A.2 Covariance matrices . . . . . . . . . . . . . . 296
- A.3 Simple linear regression . . . . . . . . . . . . 298
- A.4 Multiple linear regression (MLR) . . . . . . . . 311
- A.5 An intuitive description of principal component analysis (PCA) . . . . . . . . . . . 315

## B Appendix B. NIR spectroscopy . . . . . . . . 323
- B.1 General aspects . . . . . . . . . . . . . . . . 323
- B.2 More specific aspects . . . . . . . . . . . . . 324

## C Appendix C. Proposed calibration and classification procedures . . . . . . . . . . . 329
- C.1 Multivariate calibration . . . . . . . . . . . . 329
- C.2 Multivariate discriminant analysis . . . . . . . 332
- C.3 Cluster analysis . . . . . . . . . . . . . . . . 333

## Subject index . . . . . . . . . . . . . . . . . . . 335

# 1 Preface

The inspiration and most of the material for this book came from the Chemometric Space columns that have appeared in *NIR news* since 1991. Tormod Næs and Tomas Isaksson edited these from 1991 until the end of 1995 when Tom Fearn took over. There have been occasional guest columns over the years, but one or more of these three authors wrote most of the material. Tony Davies has been writing/editing chemometric columns since 1989 [in *Spectroscopy World* (1989–92) and *Spectroscopy Europe* (from 1992)]. His main role in this project has been to make the text as reader friendly as possible.

The Chemometric Space is aimed specifically at readers using chemometrics in the context of near infrared (NIR) spectroscopy. The techniques described are, however, much more widely applicable. Thus, although the columns were our starting point, our aim in rewriting and expanding them was to make this book a useful one for any chemometrics practitioner. Most of the examples still involve NIR spectroscopy, and some material that is specific to this application has been included, but we have tried to emphasise the generality of the approaches wherever possible.

The aim of this book is to provide a readable introduction to the disciplines of multivariate calibration and classification. We hope, as the title indicates, that the present book will serve as a useful *guide* to these important areas of science. The text is focused on the conceptual ideas behind the methods and the relationships between them. We aim at giving the reader an overview of a number of topics rather than going deeply into a few specific ones. Those who want to learn more about some of the topics can find

NIR: near infrared

*NIR news*: www.nirpublications.com

*Spectroscopy Europe*: www.spectroscopyeurope.com

Near infrared radiation is from 780 nm (12,800 cm$^{-1}$) to 2500 nm (4000 cm$^{-1}$)

Chemometrics is the use of statistical and mathematical procedures to extract information from chemical (and physical) data.

There is an introduction to matrix algebra in Appendix A

further reading in the extensive reference list at the end of the book.

The technical level of the different sections in the book varies, but mostly the mathematics is kept at a moderate level. Some knowledge of simple matrix algebra and basic statistics will, however, make the reading much easier. Our aim has, however, been to write the text in such a way that readers with less interest in the mathematical aspects can still understand the most important points.

The book is organised the following way: Chapters 3 and 4 introduce basic concepts and ideas. The most important topics are the distinction between classical and inverse calibration, the main reasons why multivariate methods are needed, the problem of multicollinearity, data compression and overfitting. Chapters 5, 6, 7 and 8 describe the most important linear calibration methods used in practice. These comprise standard and much used methods such as principal component regression and partial least squares regression, but also newer and less used techniques based on, for instance, wavelets. How to handle non-linearity problems is discussed in Chapters 9, 10, 11 and 12. Both non-linear calibration methods and methods for pre-processing the data are covered. The important problem of model validation is discussed in Chapter 13. In particular this is important for assessing the quality of a predictor and for deciding which predictor to use in practice. The next topic is outlier detection (Chapter 14), which should always be taken seriously in practical applications. How to select samples for calibration is discussed in Chapter 15 and how to monitor, maintain, update and correct calibrations are topics covered in Chapters 16 and 17. The last Chapter 18 is about classification. Many of the topics covered in other chapters of the book, for instance validation and outlier detection, are also of importance here, so several references are given to these chapters.

In the appendices, we provide some mathematical and statistical background for those readers who need a refresher course in topics such as linear regression, simple matrix algebra and principal component analysis. A short introduction to NIR spectroscopy and proposed calibration and classification procedures are also incorporated.

## 1.1 Marginal notes

You will find a number of notes in the margins throughout the book. These have been planned to give you additional information, to place information such as abbreviations where it is used to or to draw your attention to a particular point.

The following symbols have been used:

AA: abbreviations;
ⓘ: additional information;
↳ : notes or "foot"notes;
⊶: important point.

## 1.2 Acknowledgements

We would like to thank Gerry Downey for letting us use an illustration from one of his papers (Figure 18.4). During the writing of this book some of us have been financially supported by our employers. We would like to thank MATFORSK and The Agricultural University of Norway for the support. Finally, we are grateful to the publishers who have given permission to use previously published illustrations.

# 2 Introduction

This book will focus on two of the most important and widely used chemometric methodologies, multivariate calibration and multivariate classification.

Multivariate calibration is probably the area within chemometrics which has attracted the most interest so far [see, for instance, Martens and Næs (1989), Brown (1993) and Sundberg (1999)]. It is a discipline with a focus on finding relationships between one set of measurements which are easy or cheap to acquire, and other measurements, which are either expensive or labour intensive. The goal is to find good relationships such that the expensive measurements can be predicted rapidly and with high accuracy from the cheaper ones. In routine use such calibration equations can save both money and time for the user.

Multivariate classification [Mardia *et al.* (1979)] is split into two equally important areas, cluster analysis and discriminant analysis. Cluster analysis methods can be used to *find* groups in the data without any predefined class structure. Cluster analysis is highly exploratory, but can sometimes, especially at an early stage of an investigation, be very useful. Discriminant analysis is a methodology which is used for building classifiers for allocating unknown samples to one of several groups. This latter type of method has much in common with multivariate calibration. The difference lies in the fact that while multivariate calibration is used to predict continuous measurements, discriminant analysis is used to predict which class a sample belongs to, i.e. to predict a categorical variable.

In the present book we will describe the most important problems within these areas and present and

AA
PCR: principal component regression
PLS: partial least squares

discuss important solutions to them. Our aim is *not* to give a complete overview of all possible methods, but to provide a readable introduction to the area for those with little or moderate knowledge of the topics. A number of key references are given in the back of the book for those who want to study some of the topics in more detail.

As described in the preface, the main parts of the book are based on the authors' contributions to the "Chemometric Space" in the periodical *NIR news*. Therefore, most of the examples used for illustration will be taken from the area of NIR spectroscopy. Most of the methods to be discussed below are, however, applicable far outside the area of NIR spectroscopy. Readers who are not familiar with NIR spectroscopy may find the short introductory treatment given in Appendix B useful.

## 2.1 Basic problems and their solutions

The most important problems arising in calibration are:
- the non-selectivity problem
- the collinearity problem
- the non-linearity problem
- calibration data selection
- the outlier problem

The first of these points is the most basic. In spectroscopy, for example, it is often difficult in practice to find selective wavelengths for the chemical constituents in the samples. For instance, when calibrating for protein content in wheat, no single NIR wavelength provides sufficient information. The absorbances at all wavelengths are affected by a number of the chemical and physical properties of the sample. The selectivity problem can be solved in spectroscopy by using several wavelengths in the calibration equation. This is called multivariate calibration.

Variables are collinear if there are high correlations or other near or exact linear relations among them. This frequently occurs in modern chemistry due to the easy access to a large number of strongly related measurement variables. Standard regression techniques based on least squares estimation give very unstable and unreliable regression coefficients and predictions when used with collinear variables, so other methods must be used. A number of the best-known calibration techniques, for instance principal component regression (PCR) and partial least squares (PLS) regression, were developed for solving the collinearity problem.

The next problem has to do with modelling, i.e. finding the best model for the relationship between spectral or other inexpensive measurements and the reference chemistry. Usually, a linear model is good enough, but in some cases non-linear alternatives can give substantial improvement.

In all practical calibration situations the model contains unknown constants, parameters, that need to be estimated using data on a number of calibration samples. The quality of the resulting calibration equation can depend strongly on the number of calibration samples used and also on how the calibration samples were selected. The samples should span the natural range of variability as well as possible in order to give a predictor which can be used safely for all types of unknown samples.

The last problem has to do with errors or unexpected variability in particular observations, so-called outliers. These are observations or variables that in some way are different from the rest of the data. There may be several reasons for such outliers. Instrumental errors or drift or samples from other populations than expected are typical examples. Outliers can lead to

**2.1 Basic problems and their solutions**

grossly erroneous results if they pass unnoticed, so it is very important to have methods to detect them.

All the five problems listed above were explained in a calibration context. All the problems except the non-linearity problem are, however, equally important for classification situations.

The present book is devoted to explaining and discussing useful methods for solving these problems. The methods to be focussed on in this book are intended to be used by others than specialist chemometricians. For such users it is important that the techniques are

▶▶ simple to use
▶▶ simple to understand
▶▶ simple to modify

In the present text, which is meant as an introduction for non-experts, we have therefore tried to emphasise the techniques that satisfy these requirements best.

The book has no direct link to specific software packages. Many of the basic problems can be solved by using standard chemometric software packages such as Unscrambler, SIMCA, Sirius, PLS-toolbox and Pirouette. For the more sophisticated methods, however, one will need either to use more comprehensive statistical packages such as SAS, S+ etc. For some of the most recent methods, one will need to write special programs in, for instance, MATLAB, SAS or S+.

In this book we will use standard nomenclature and symbols as far as possible. Upper case bold face letters (**X**, **A** etc.) will be used to denote matrices. Lower case bold face letters (**x**, **y** etc.) will be used to denote vectors and both upper and lower case italics will be used to denote scalars ($A, K, x, y$ etc.). Empirical estimates of model parameters will be denoted by a hat, ^. An estimate of a parameter $b$ is thus denoted

Unscrambler:
www.camo.com
PLS-Toolbox:
www.eigenvector.com
SIMCA:
www.umetrics.com
Pirouette:
www.infometrics.com
Sirius: www.prs.no
SAS: www.sas.com
S+: www.mathsoft.com
Matlab:
www.mathworks.com

*Matrices:* Upper case bold face letters (**X**, **A**)
*Vectors:* lower case bold face letters (**x**, **y**)
*Scalars:* italics ($A, K, x, y$)
*Empirical estimates of model parameters:* hat, ^

by $\hat{b}$. More information about symbols, basic concepts and principles can be found in Appendix A.

**2.1** Basic problems and their solutions

# 3 Univariate calibration and the need for multivariate methods

Although the present book is about multivariate methods, we will first give a brief description of how to perform calibration in the univariate case, i.e. when we have only one variable ($x$) to measure and one variable ($y$) to predict. This section also serves as an introduction to the terminology used in the rest of this book. At the end of the chapter we will indicate the shortcomings of univariate methodology and discuss some reasons why multivariate techniques are often needed.

LS: least squares
MLR: multiple linear regression
MSC: multiplicative scatter correction
MSE: mean square error
NIR: near infrared
PLS: partial least squares

## 3.1 Classical vs inverse calibration

Suppose we have a set of calibration data with a rapid measurement $x$ and a reference measurement $y$ taken on each of $N$ samples. Let us assume that the relationship between $x$ and $y$ is reasonably linear. There are two obvious ways to fit a linear calibration line for predicting $y$ from $x$:

(1) fit the model $y = b_0 + b_1 x + error$ to the $N$ data points by least squares (LS) and use the equation $\hat{y} = \hat{b}_0 + \hat{b}_1 x$ directly for prediction.
(2) fit the model $x = a_0 + a_1 y + error$ by least squares (LS), invert the equation to give $\hat{y} = -(\hat{a}_0 / \hat{a}_1) + (1 / \hat{a}_1) x$, and use this for prediction.

Beer's law
Bouguer (1729) and Lambert (1760) independently proposed that in a uniform, transparent medium the light intensity, $I$, decreases exponentially with increasing path length, $L$, from the incident value, $I_0$. In 1852 Beer proposed that absorption of light is proportional to the concentration, $c$, of the absorbing species. These two ideas are combined in the equation.
$A = \log(I_0/I) = kLc$
where $A$ is the absorption and k is the absorption coefficient. It is commonly called "Beer's Law"; occasionally you will see "Beer–Lambert". Bouguer is mentioned very infrequently!

$R^2$
The correlation coefficient $R$ measures how well the prediction agress with the reference. $R$ has values from −1.0 to 1.0, $R^2$ has values between 0 and 1.0; a value of 0 indicates no correlation, 1.0 perfect linear relationship.

Rather confusingly (1) is referred to in most of the literature as "inverse calibration" and (2) as "classical calibration" [see Martens and Næs (1989), Halperin (1970), Shukla (1972) and Oman (1985)]. The reason for this is historical; method (2) assumes a linear model for how the spectral or other rapid reading is affected by the chemical concentration. In spectroscopy this linear model is called Beer's model or Beer's law. Using this model for calibration is traditional in chemistry and it is therefore called the classical approach. The other approach was introduced in statistics much later and is called the inverse method because it does the opposite of the classical one.

In Figure 3.1, the less steep of the two lines corresponds to the inverse procedure (1), the steeper one to the classical one (2). Both lines pass through the point $(\bar{x}, \bar{y})$, whose co-ordinates are the mean of the $x$ and $y$ values of the calibration data. The figure shows the two predictions of $y$ for an observed $x$ equal to 17. The prediction using the inverse line (1) is closer to the mean $\bar{y}$ than is the prediction using the classical line. This will be true for all values of $y$: the inverse calibration "shrinks" predictions towards the mean compared with the classical calibration, the so-called least squares effect. How large this effect is depends on how far apart the lines are, which in turn depends on how good the fit is. For a tight calibration, the effect will be negligible. For calibrations with moderate to low $R^2$ (e.g. 0.9 or less) it may be appreciable, especially for values far from the mean.

Shrinking towards the mean is a good idea if the reference value you are trying to predict is close to the mean, less good if it is far away. One way of trying to make this more precise is to study the mean square error ($MSE$) properties (see Chapter 13) of the two methods. The $MSE$ is the expectation of the squared difference between $\hat{y}$ and $y$ (expectation can be interpreted as the average taken over the population of fu-

3.1 Classical vs inverse calibration

# Univariate calibration and the need for multivariate methods

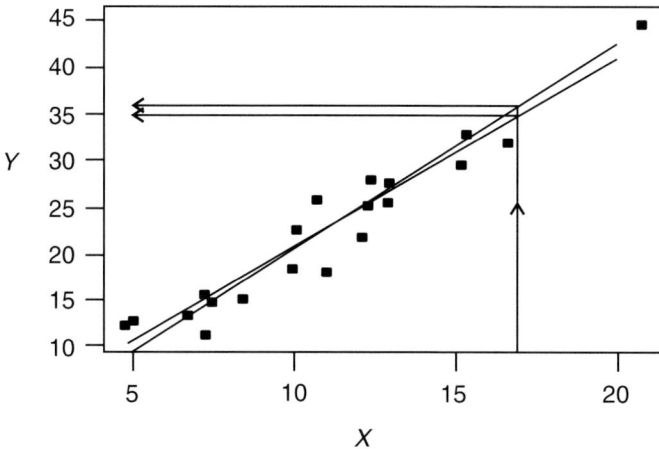

Figure 3.1. Calibration lines from the inverse (flatter line) and classical method (steeper line) for a set of 20 calibration samples. To predict the reference value for an unknown with rapid measurement 17, we find where the vertical line from 17 intersects both lines, and read across to the reference axis as shown.

ture samples). It turns out that the inverse method does indeed have better mean squared error properties than the classical method for reference values close to the mean, and poorer properties for values far away (see Figure 3.2). The crossover point is, roughly, at $1.5 s_y$ [see, for example, Fearn (1992)], where $s_y$ is the standard deviation of the calibration reference values.

If the calibration reference values are representative of the population we wish to predict, this will ensure that most future unknowns lie in the region where the inverse method is better, and this method is thus to be preferred. This will be true whether the calibration samples have been randomly or systematically chosen. If calibration samples have been chosen systematically to give a more uniform distribution, this will increase $s_y$ and make it even more likely that future randomly chosen samples are better predicted with the inverse than the classical method. A possible exception to this preference for the inverse method is

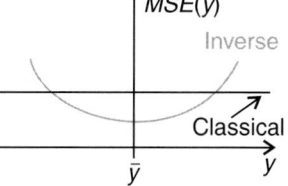

Figure 3.2. $MSE(\hat{y})[= E(\hat{y} - y)^2]$ properties of the two regression methods, the inverse and the classical. The $MSE(\hat{y})$ is here presented as a function of the y-value of the sample to be predicted. Note that this is different from in Chapter 13 where the average MSE taken over all y-values is in focus. The curves here refer to the situation where the Beer's law parameters $a_0$ and $a_1$ are known. For practical situations, the true MSE curves will only approximate these idealised curves.

## 3.1 Classical vs inverse calibration

when prediction errors for extreme samples are the ones that really matter, which might be the case in some quality control situations. If calibration samples represent a random selection from some population, then it can be shown that the inverse method is optimal in the sense that it minimises mean square error of predictions averaged over future unknown samples from the same population. We refer to Sundberg (1985) for more information.

Interestingly, a flat distribution of calibration samples has a second effect. For a given range, the closer the data points are to the ends of the range, the closer together will be the two lines (and then the two predictors). For a flat distribution of points the lines will be closer together than they are for a random sample with a concentration of points in the middle.

## 3.2 Why are multivariate methods needed?

If all problems could be solved by modelling the relationship between two variables, this book would be a slim volume. There are, however, many examples where several predictor variables used in combination can give dramatically better results than any of the individual predictors used alone. One rich source of such examples is NIR spectroscopy (see Appendix B for a short introduction).

When the peaks in a spectrum (due to different absorbers) overlap, it will not usually be possible to use absorbance at a single wavelength to predict the concentration of one of the absorbers. This problem is often called the *selectivity* problem. The problem and its solution by multivariate techniques are illustrated in Figure 3.3. This example is based on NIR measurements of 103 samples of homogenised pork and beef. The spectra are measured in diffuse transmittance mode, scanned from 850 to 1050 nm in 2 nm steps

## Univariate calibration and the need for multivariate methods

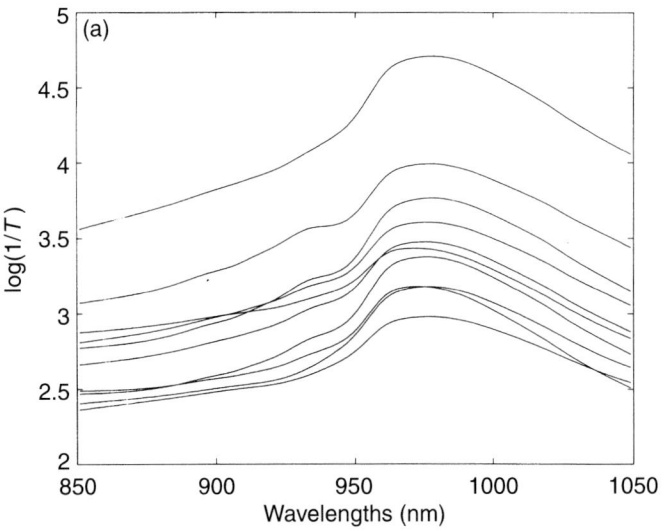

**Figure 3.3.** Illustration of the selectivity problem in NIR spectroscopy. Pork and beef data. (a) log(1/*T*) NIR spectra of 10 selected samples.

(giving 100 *x*-variables). Ten spectra are presented in Figure 3.3(a). The fat content for each sample was measured by a laboratory "wet chemistry" method. Further description of this dataset can be found in Næs and Isaksson (1992b).

The best predictions of fat content using only one spectral variable were obtained at 940 nm, corresponding to the third overtone absorbance band of the $-CH_2-$ group. The measured and predicted results are plotted in Figure 3.3(b). The corresponding cross-validated (see Chapter 13) correlation between predicted and measured fat content was equal to 0.23. As can be seen, even the best possible prediction results obtained by a univariate method are very poor.

In such cases, it is often advantageous to combine information from several or even all the spectral variables. These *multivariate* techniques also have a number of other advantages, which will be illustrated later in the book (see, for instance, Chapter 14). A

Third overtone
See the appendix on NIR spectroscopy if you are not familiar with this term.

### 3.2 Why are multivariate methods needed?

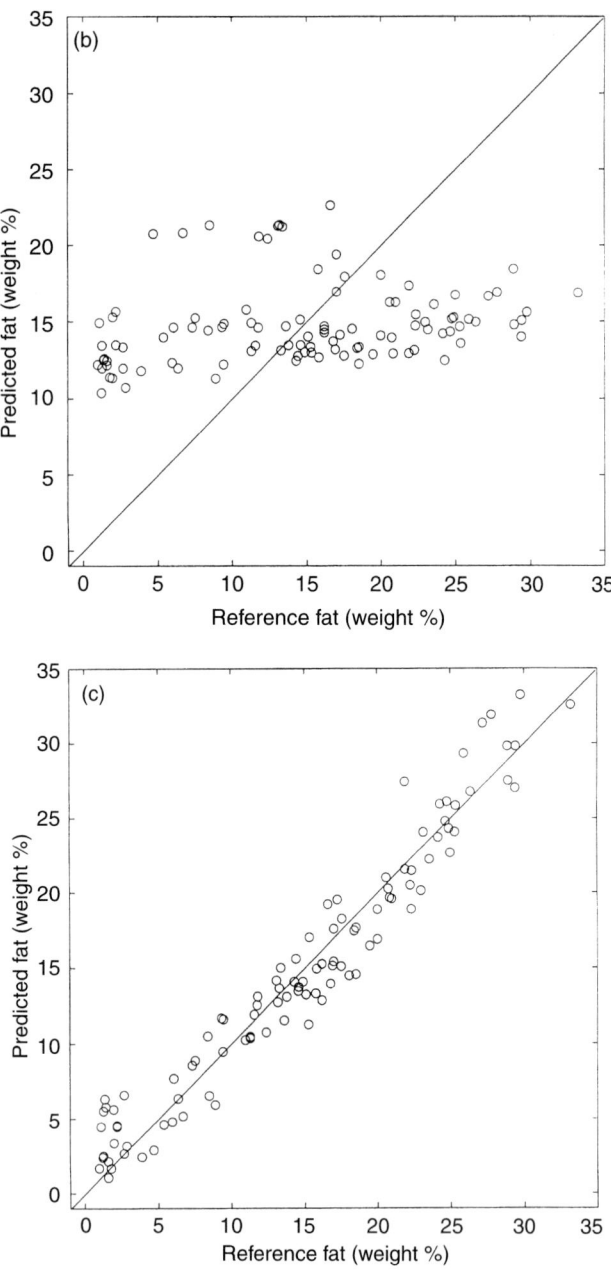

Figure 3.3. Illustration of the selectivity problem in NIR spectroscopy. Pork and beef data. (b) Reference method fat vs predicted fat, using one variable (940 nm) as a predictor (correlation = 0.23). (c) Reference method fat vs predicted fat using a multivariate calibration method (PLS) based on all variables from the spectral data (correlation = 0.97).

## 3.2 Why are multivariate methods needed?

multivariate PLS regression (see Chapter 5) was applied to the same data and gave a correlation of 0.97. The measured and predicted fat concentrations are presented in Figure 3.3(c).

For this dataset, both the univariate and multivariate calibrations were improved by using multiplicative scatter correction (MSC, see Chapter 10). The univariate correlation improved from 0.23 to 0.77 and the multivariate from 0.97 to 0.99. This shows that some but not all the lack of selectivity comes from the scatter effect (see Appendix B). The MSC reduces the problem, but does not solve it. The rest of the problem comes from overlap of the individual spectra for the constituents in the samples.

All the ideas and results concerning inverse vs classical calibration extend to multivariate **x**. Multivariate classical methods based on Beer's law are discussed in, for instance, Martens and Næs (1989), but they will not be pursued further here. Instead we focus on the inverse approach, which is usually to be preferred. The standard multiple linear regression (MLR) approach [see, for instance, Appendix A, Weisberg (1985) and Mark (1991) for details] is the simplest way of performing an inverse multivariate calibration. It is a straightforward generalisation of the univariate inverse method based on least squares fitting of $y$ to **x**. Due to a number of problems, for instance collinearity and the danger of outliers, other and more sophisticated approaches are usually to be preferred. For the rest of this book, a number of important aspects related to these other methods of analysis will be the main focus.

**3.2** Why are multivariate methods needed?

# 4 Multicollinearity and the need for data compression

## 4.1 Multicollinearity

As was mentioned above, multivariate calibration can usually not be handled by the standard MLR method based on the least squares (LS) criterion. One of the reasons for this is that the number of available samples is often much smaller than the number of $x$-variables. This leads to exact linear relationships, so-called exact multicollinearity,[*] among the variables in the matrix $\mathbf{X}$. The matrix $\mathbf{X}^t\mathbf{X}$ in the MLR equation is then singular (non-invertible, see Appendix A) and the LS solution becomes non-unique. But even if the number of samples is increased, MLR will often give a bad predictor. This is due to the so-called near-multicollinearity often found among, for example, spectral measurements. This means that some of the variables can be written approximately as linear functions of other variables. The predictor will in such cases be mathematically unique, but it may be very unstable and lead to poor prediction performance, as will be illustrated below [see also Weisberg (1985) and Martens and Næs (1989)].

Sometimes, multicollinearity is just called collinearity

**Aa**
*CN:* condition number
*LS:* least squares
*MLR:* multiple linear regression
*PCR:* principal component regression
*PLS:* partial least squares
*VIF:* variance inflation factor

In the rest of this chapter we will assume that the multivariate linear model

$$y = b_0 + \sum_{k=1}^{K} b_k x_k + f \qquad (4.1)$$

gives an adequate fit to the data. If, for instance, Beer's law applies, then this is a reasonable assumption. We let $N$ be the number of samples available for

*K:* number of x-variables
*N:* number of samples

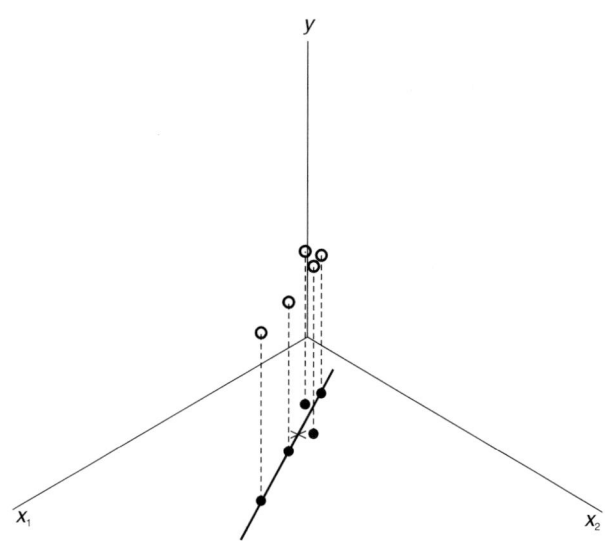

Figure 4.1. Graphical illustration of a situation where *y* is modelled as a function of two multicollinear *x*-variables. A linear model for such a relationship corresponds to a plane in the three-dimensional space. High correlation among the *x*-variables, •, is shown by their closeness to the straight line in the $x_1$–$x_2$ plane. The cross on the straight line corresponds to the average in *x*-space. Reproduced with permission from H. Martens and T. Næs, *Multivariate Calibration* (1989). © John Wiley & Sons Limited.

Strictly **y** is a vector, but a vector is just a special type of matrix, and we often blur the distinction.

calibration and we denote the matrices of *x*- and *y*-values by **X** and **y**. Note that *K* is the symbol used for the number of *x*-variables in the equation.

For only two *x*-variables, near-multicollinearity is identical to a high correlation among them. The multicollinearity (or just collinearity) problem is illustrated for such a situation in Figure 4.1. With only two predictor variables, estimating the regression coefficients in Equation (4.1) corresponds to fitting a plane to the data. From the figure it is clear that orthogonal to the straight line (first principal component, see Chapter 5) drawn in the $x_1$–$x_2$ plane, there is very little variability and therefore limited information about the regression plane. The LS solution will therefore be very unstable in this direction [see also

Næs and Martens (1988)]. A small disturbance in one of the *y*-values can lead to a substantial change in the plane. The result of this instability is often unreliable prediction results.

There are a number of ways that near- or exact-multicollinearity can be detected or diagnosed. One of them is simply to look for correlated variables as in Figure 4.1. This is, however, not always good enough since some types of multicollinearities may be more complex. For instance, one of the variables may be approximated by a linear function of four other variables without any two of the variables being highly correlated.

Two much used techniques in statistics for detecting near-multicollinearity are the variance inflation factor (*VIF*) and the condition number (*CN*). The *VIF* is defined for each variable *k* and can be written as

$$VIF_k = 1/(1 - R_k^2) \qquad (4.2)$$

where $R_k^2$ is the squared correlation coefficient between $x_k$ and the linear regression predictor of $x_k$ based on all the other *x*-variables [see Weisberg (1985) and Belsley *et al.* (1980)]. The *VIF* measures directly how much the variance of each coefficient is inflated as compared to a situation with uncorrelated *x*-variables (orthogonal design). For an orthogonal design [Box *et al.* (1978)], all *VIF*s are equal to 1. The larger the *VIF* is, the more serious is the collinearity problem. An important advantage of using *VIF* is that it points out exactly for which variables the multicollinearity is a problem. For exact collinearity, at least one $R_k^2$ becomes equal to 1 and the corresponding *VIF*s will be infinite.

The condition number (*CN*) is defined by

$$\kappa = (\hat{\lambda}_1 / \hat{\lambda}_K)^{1/2} \qquad (4.3)$$

where $\hat{\lambda}_1$ is the largest and $\hat{\lambda}_K$ the smallest eigenvalue of the empirical covariance matrix of **x**,  Eigenvalue See Appendix A

**4.1 Multicollinearity**

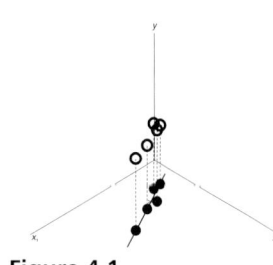

Figure 4.1

$\hat{\Sigma} = (\mathbf{X} - \mathbf{1}\bar{\mathbf{x}}^t)^t (\mathbf{X} - \mathbf{1}\bar{\mathbf{x}}^t)/(N-1)$ [see Weisberg (1985) and Appendix A]. The condition number can be interpreted as the ratio of the standard deviation along the axis with the largest variability relative to the standard deviation along the axis with the smallest variability. The situation illustrated in Figure 4.1 is typical where the variability along the axis with the largest variability (solid line) is much larger than the variability orthogonal to it. It will therefore result in a large $\kappa$ value. Rules such as declaring collinearity a problem if $\kappa \geq 30$ have been put forward, but have little theoretical basis. Note that in the case of exact collinearity, the smallest eigenvalue will be equal to 0 and the *CN* is not defined.

An illustration of the use of *VIF* and *CN* is given in the example illustrated by Figure 5.2. Eigenvectors and eigenvalues are discussed further in Chapters 5 and 6 [see also Mardia *et al.* (1979) and Appendix A].

Both these tools are useful as a first test of collinearity, but the practical effect of the collinearity in calibration can only be checked by looking at the accuracy of the results predicted by the final calibration. This is usually done by empirical validation procedures and will be the topic of Chapter 13.

## 4.2   Data compression

Many ways of solving the multicollinearity problem exist, but in the chemometric literature two approaches have become particularly popular. These are the methods which use MLR for a few carefully selected variables, and the methods which regress *y* onto a few linear combinations (components or factors) of the original *x*-variables. Both these ideas of data compression are illustrated in Figure 4.2. For the variable selection techniques [Figure 4.2(a)] some of the variables (here $x_1$ and $x_3$) are deleted before regression, usually resulting in a more stable and reliable

# Multicollinearity and the need for data compression

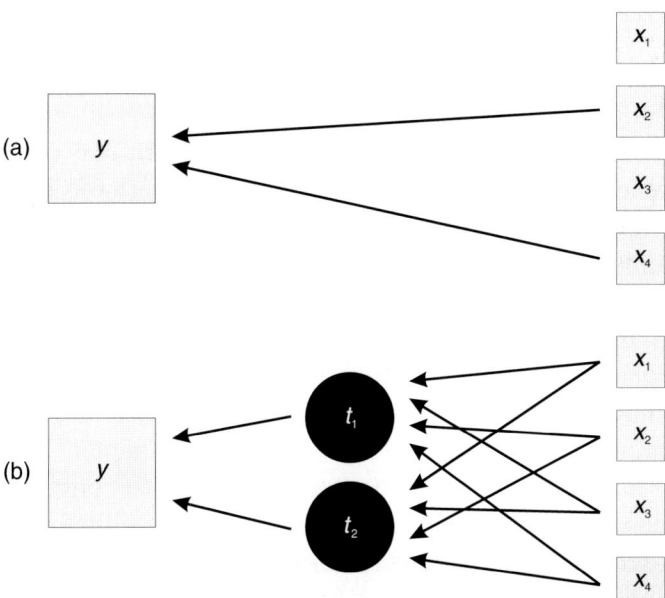

Figure 4.2. Conceptual illustration of the difference between methods for solving the multicollinearity problem. (a) indicates that variable selection deletes some of the variables from the model. (b) shows how all x-variables are transformed into linear combinations $t_1$ and $t_2$, which are related to y in a regression equation.

 PCR and PLS regression are both based on the idea in (b).

calibration equation based on the remaining, less multicollinear, variables. For the other class of methods, all variables [Figure 4.2(b)] are used in the regression, but the arrows indicate that the explanatory variables are compressed onto a few linear combinations (ts) of the original variables before regression. In both cases, the multicollinearity is solved and the prediction equation obtained is stable and reliable. Combinations of the two strategies have also been developed [see, for example, Brown (1992), Martens and Martens (2000, 2001) and Westad and Martens (2000)]. As will be discussed below, these data compression methods are also useful for better interpretation of the data.

## 4.2 Data compression

Several ways exist to select useful variables. The two which are most frequently used in NIR are the so-called *forward* selection and *best subset* selection [Weisberg (1985)]. The former starts with one variable, then incorporates one additional variable at a time until the desired number is reached. The best subset method compares the performance of all subsets of variables, sometimes with a limit on the maximum number, and ends up with a suggested best subset. Due to the high number of comparisons, this strategy is quite time consuming and can also be quite prone to overfitting (see below and Chapter 13). Among the data compression techniques that construct new variables, PLS regression and PCR are the most used methods, but other useful methods also exist.

## 4.3 Overfitting versus underfitting

An important problem for all the data compression methods is selection of the optimal number of variables or components to use. If too many components are used, too much of the redundancy in the $x$-variables is used and the solution becomes overfitted. The equation will be very data dependent and will give poor prediction results. Using too few components is called underfitting and means that the model is not large enough to capture the important variability in the data. These two important phenomena are illustrated in Figure 4.3. As we see, there are two effects that apply in this case, estimation error and model error. As the number of variables or components increases, the model error decreases, since more of the $x$-variability is modelled, while the estimation error increases due to the increased number of parameters to be estimated. The optimal value is usually found in between the two extremes. An important problem in practice is to find this optimum. In Chapter 13 this problem will be treated extensively.

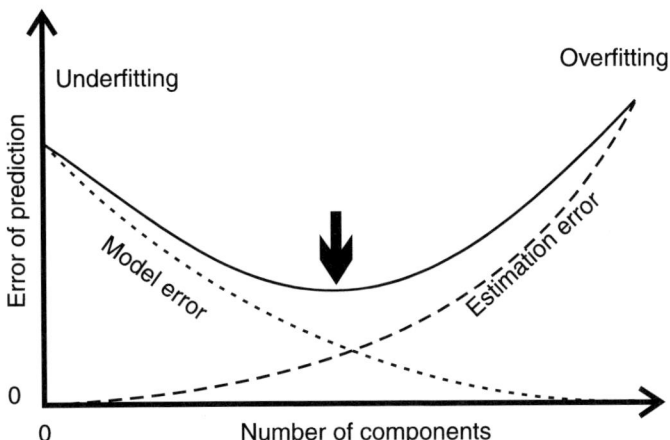

Figure 4.3. The estimation error (– – –) and model error (· · ·) contribute to the prediction error (———). If too large a model is used, an overfitted solution is the result. The opposite is called underfitting. Usually a medium size model is to be preferred. Reproduced with permission from H. Martens and T. Næs, *Multivariate Calibration* (1989). © John Wiley & Sons Limited.

The overfitting effect is strongly dependent on the number of samples used [Martens and Næs (1989)]. The more samples that are used, the more precise are the parameter estimates and thus also the predictions. Therefore, the more samples that are used, the less important is the overfitting effect. In practice, however, it must always be taken seriously.

# 5 Data compression by PCR and PLS

This chapter will be devoted to two widely used methods, PLS and PCR.

## 5.1 General model structure for PCR and PLS

The idea behind these two methods [see Figure 4.2 (b)] is to find a few linear combinations (components or factors) of the original *x*-values and to use only these linear combinations in the regression equation [Wold *et al.* (1984), Martens and Næs (1989), Martens and Martens (2001)]. In this way irrelevant and unstable information is discarded and only the most relevant part of the *x*-variation is used for regression. Experience and theory have shown that such approaches often give improved prediction results. The collinearity problem is solved and more stable regression equations and predictions are obtained. Since all variables are projected down to a few linear combinations, simple plotting techniques can also be used. This can be important for interpretation, as will be demonstrated below.

The data matrices used as input to PCR and PLS regression are denoted by **X** and **y**. They are here both assumed to be mean-centred, i.e. the mean of each variable has been subtracted from the original measurements. The model structure for both methods is given by the two equations (see also Figure 5.1)

$$\mathbf{X} = \mathbf{TP}^t + \mathbf{E}$$
$$\mathbf{y} = \mathbf{Tq} + \mathbf{f} \tag{5.1}$$

**AA**
*CN*: condition number
*CR*: continuum regression
*LS*: least squares
*MLR*: multiple linear regression
*PCA*: principal component analysis
*PCR*: principal component regression
*PLS*: partial least squares
*PLS2*: partial least squares with multiple *y*-variables
*RMSEP*: root mean square error of prediction
*RR*: ridge regression
*VIF*: variance inflation factor

Strictly **y** is a vector, but a vector is just a special type of matrix, and we often blur the distinction.

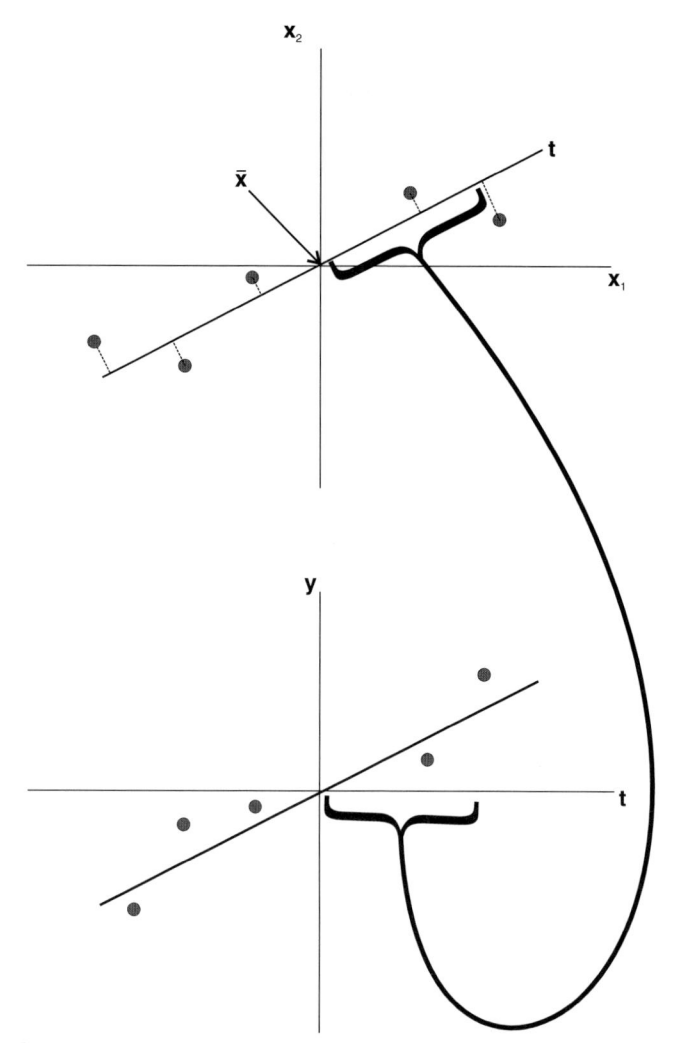

**Figure 5.1.** Geometrical illustration of the model structure used for the methods PCR and PLS. The information in X is first compressed down to a few components t before these components are used as independent variables in a regression equation with y as dependent variable. The two parts of the illustration correspond to the two equations in (5.1). The first is data compression in x-space, the other is regression based on the compressed components.

The $A$ columns of the matrix **T** may be thought of as a small set of underlying or latent variables responsible for the systematic variation in both **X** and **y**.

The matrix **P** and vector **q** are called loadings and describe how the variables in **T** relate to the original data matrices **X** and **y**. The matrix **E** and the vector **f** are called residuals and represent the noise or irrelevant variability in **X** and **y**, respectively. Sometimes the model (5.1) is called a bilinear model, since it is linear in both loadings and scores.

During estimation of model parameters, the two methods use different criteria for the computation of the scores matrix **T**. These two criteria will be defined and discussed below. In both cases the variables in **T** are linear combinations of the original variables in **X**. The rest of the computations are identical; the loadings in **P** and **q** are estimated by regressing **X** and **y** onto **T**, and the residual matrices are found by subtracting the estimated versions of **TP**′ and **Tq** from **X** and **y**, respectively. The estimated model parameters can, in both cases, be combined into the regression vector $\hat{\mathbf{b}}$ to be used in the prediction equation

$$\hat{y} = \hat{b}_0 + \mathbf{x}'\hat{\mathbf{b}} \tag{5.2}$$

When using mean-centred $x$-variables in the model, the intercept $\hat{b}_0$ becomes equal to $\bar{y}$, the mean of the original $y$-values for the calibration samples. Note that in equation (5.2), $\hat{y}$ is the predicted value of $y$ without mean centring.

Below we discuss how computation of $\hat{\mathbf{b}}$ is carried out for both PCR and PLS. Prediction can, however, also be performed directly using the equations in (5.1) without going via formula (5.2) [see Næs and Martens (1988)].

For both PCR and PLS, hats ^ will be put on the estimated quantities, $\hat{\mathbf{P}}$, $\hat{\mathbf{T}}$ etc. This is done in order to emphasise that they are estimates of the model quantities in equation (5.1). Since $\hat{\mathbf{P}}$ and $\hat{\mathbf{T}}$ for the PCR method are defined as regular principal component loadings and scores, respectively, which are usually presented without hats [Mardia et al. (1979)], these particular quantities could have been left without hats. For consistency, however, and to point out explicitly that they are computed from data, the hats will be used here for PCR as well.

## 5.2 Standardisation of *x*-variables

Neither of the PCR or PLS methods is invariant to changes of the scale of the $x$-variables. If the units of some of the variables are changed, the estimated parameters and the regression vector will also be changed. This may become an important issue if *x*-variables used in the model are measured on very different scales, as may be the case if measurements from different instruments are collected together in **X**.

In such cases it may be wise to standardise the variables before calibration. This means that each variable is divided by its standard deviation (taken over the calibration set), resulting in variables all having a standard deviation equal to 1. A possible pitfall with this strategy is that variables with small variance and with a low signal-to-noise ratio may become more important than they should be. Therefore, uncritical standardisation of the variables without considering the signal-to-noise ratios should be avoided.

In NIR spectroscopy and related methodologies, the measurement units for the variables are comparable across the whole spectrum. In such cases, standardisation is usually not performed. Standardisation may also sometimes preclude interpretation of the results.

## 5.3 Principal component regression (PCR)

The PCR method is based on the basic concept of principal component analysis (PCA). Those readers who need an introduction to this important method are referred to Mardia *et al.* (1979) (see also Appendix A5). An example of using the method for interpretation is given in Section 6.4.

For PCR, the estimated scores matrix $\hat{\mathbf{T}}$ consists of the $A$ most dominating principal components of **X**. These components are linear combinations of **X** determined by their ability to account for variability in **X**. The first principal component, $\hat{\mathbf{t}}_1$, is computed as the linear combination of the original *x*-variables with the highest possible variance. The vector defining the linear combination is scaled to have length 1 and is denoted by $\hat{\mathbf{p}}_1$. The second component, $\hat{\mathbf{t}}_2$, is then defined the same way, but under the constraint that it is uncorrelated with $\hat{\mathbf{t}}_1$. The second direction vector is also of unit length and is denoted by $\hat{\mathbf{p}}_2$. It can be shown that this is orthogonal to $\hat{\mathbf{p}}_1$. The process continues until the desired number of components, $A$, has been extracted. See Chapter 13 for information about how an appropriate $A$ can be determined. In principle, the process can continue until there is no variability left in $\mathbf{X}$. If the number of samples is larger than the number of variables, the maximum number of components that may be computed is equal to the number of variables $K$.

> Alternatively, the constraint that $\hat{\mathbf{p}}_1$ be orthogonal to $\hat{\mathbf{p}}_2$ may be imposed. This results in the same solution, which has the property that $\hat{\mathbf{t}}_1$ and $\hat{\mathbf{t}}_2$ are uncorrelated.

The matrix consisting of the $A$ most dominating principal component scores is denoted by $\hat{\mathbf{T}}$ and the corresponding matrix of loadings is denoted by $\hat{\mathbf{P}}$. Sometimes a subscript $A$ is used for both $\hat{\mathbf{T}}$ and $\hat{\mathbf{P}}$ to indicate the number of columns, but this is avoided here. It can be shown that with these definitions the centred matrix $\mathbf{X}$ can be written as

$$\mathbf{X} = \hat{\mathbf{T}}\hat{\mathbf{P}}' + \hat{\mathbf{E}} \qquad (5.3)$$

showing that $\mathbf{X}$ can be approximated by a product of the $A$ first scores and their corresponding loadings. It can also be shown that no other $A$-dimensional approximation gives a better fit to $\mathbf{X}$, i.e. one for which the sum of squares of elements of $\hat{\mathbf{E}}$ is smaller. Note that equation (5.3) is an estimated version of the first model equation in (5.1).

The principal components can easily be computed using the eigenvector decomposition of the cross-product matrix $\mathbf{X}'\mathbf{X}$. Since the $\mathbf{X}$-matrix is centred, this matrix is identical to $N-1$ times the empirical covariance matrix defined in Appendix A. The columns of the $\hat{\mathbf{P}}$-matrix are the unit length eigenvectors of this matrix. The scores matrix $\hat{\mathbf{T}}$ can most easily be found by regressing $\mathbf{X}$ onto $\hat{\mathbf{P}}$, giving the solution $\hat{\mathbf{T}} = \mathbf{X}\hat{\mathbf{P}}$. The eigenvalues of the cross-product matrix are identical to the sums of squares of the columns of $\hat{\mathbf{T}}$. The $A$ first principal component scores correspond to the $A$ eigenvectors with the largest eigenvalues.

The next step is to use the matrix $\hat{\mathbf{T}}$ in the regression equation instead of the original variables in $\mathbf{X}$. The regression model can be written as

$$\mathbf{y} = \hat{\mathbf{T}}\mathbf{q} + \mathbf{f} \qquad (5.4)$$

and the regression coefficients in $\mathbf{q}$ are estimated by regular least squares. With $A$ equal to its maximal value $K$, the equation (5.4) becomes equivalent to the full regression equation and the PCR predictor becomes identical to the MLR predictor.

The idea behind the PCR method is to remove the $\mathbf{X}$-dimensions with the least variability from the regression. As was discussed in Chapter 4, these are the main reasons for the instability of the predictions. Therefore, the principal components corresponding to the smallest eigenvalues are omitted from the regression equation. This intuitive idea is also supported by comparing theoretical formulae for the prediction ability of PCR and LS regression [see, for example, Næs and Martens (1988)]. These results show clearly that PCR can give substantially more stable regression coefficients and better predictions than ordinary LS regression.

Predicting $y$ for new samples can be done in two equivalent ways. One possibility is to compute $\hat{\mathbf{t}}$ for each sample using the formula $\hat{\mathbf{t}}' = \mathbf{x}'\hat{\mathbf{P}}$ (centred $\mathbf{x}$), and then to use this $\hat{\mathbf{t}}$ in the prediction equation $\hat{y} = \bar{y} + \hat{\mathbf{t}}'\hat{\mathbf{q}}$ corresponding to equation (5.4). The other way is to use the linear predictor $\hat{y} = \bar{y} + \mathbf{x}'\hat{\mathbf{b}}$ directly where the regression coefficient vector $\hat{\mathbf{b}}$ is computed as

$$\hat{\mathbf{b}} = \hat{\mathbf{P}}\hat{\mathbf{q}} \qquad (5.5)$$

Note that the intercept in both cases is equal to $\bar{y}$ since the **X**-matrix is centred.

It should be mentioned that there are other ways to select principal components for regression. Some authors advocate the use of t-tests (see Appendix A). The idea behind this approach is that PCR, as defined above, selects components only according to their ability to account for variability in **X** and without using information about $y$. One is then accepting the risk that some of the components have little relevance for predicting $y$. Using t-tests is one such way of testing for relevance of the components, which in some cases may lead to improvements. In other cases, the opposite can also happen. Our preference for most applications in chemistry is selection according to size of the eigenvalue. For a broader discussion of this point we refer to Næs and Martens (1988) and Joliffe (1986).

## 5.4 Partial least squares (PLS) regression

One of the reasons for the development of the PLS regression method was to avoid the dilemma in PCR of deciding which components to use in the regression equation. Instead of using selected principal components in $\hat{\mathbf{T}}$, PLS uses factors determined by employing both **X** and **y** in the estimation directly. For PLS regression each component is obtained by *maximising the covariance* between **y** and all possible linear functions of **X** [see, for example, Frank (1987) and Høskuldsson (1988)]. This leads to components, which are more directly related to variability in **y** than are the principal components.

The direction of the first PLS component, obtained by maximising the covariance criterion, is denoted by $\hat{\mathbf{w}}_1$. This is a unit length vector and is often called the first *loading weight* vector. The scores along this axis are computed as $\hat{\mathbf{t}}_1 = \mathbf{X}\hat{\mathbf{w}}_1$. All variables in **X** are then regressed onto $\hat{\mathbf{t}}_1$ in order to obtain the loading vector $\hat{\mathbf{p}}_1$. The regression coefficient $\hat{q}_1$ is obtained similarly by regressing $\mathbf{y}$ onto $\hat{\mathbf{t}}_1$. The product of $\hat{\mathbf{t}}_1$ and $\hat{\mathbf{p}}_1$ is then subtracted from $\mathbf{X}$, and $\hat{\mathbf{t}}_1\hat{q}_1$ is subtracted from $\mathbf{y}$. The second direction is found in the same way as the first, but using the residuals after subtraction of the first component instead of the original data. The process is continued in the same way until the desired number of components, $A$, is extracted. If $N > K$, the process can continue until $A = K$. In this case PLS is, as was PCR, identical to MLR.

Note that for PLS, the loading weights are not equal to the loadings $\hat{\mathbf{P}}$. For PCR, however, only one set of loadings was required. It can be shown that the PLS loading weight column vectors are orthogonal to each other, while the PLS loading vectors are not. The columns of the PLS scores matrix $\hat{\mathbf{T}}$ are orthogonal. The matrix $\hat{\mathbf{P}}$ and vector $\hat{\mathbf{q}}$ can therefore, as for PCR, be obtained by regressing $\mathbf{X}$ and $\mathbf{y}$ onto the final PLS scores matrix $\hat{\mathbf{T}}$.

The regression coefficient vector used in the linear PLS predictor can be computed using the equation

$$\hat{\mathbf{b}} = \hat{\mathbf{W}}(\hat{\mathbf{P}}'\hat{\mathbf{W}})^{-1}\hat{\mathbf{q}} \qquad (5.6)$$

where the $\hat{\mathbf{W}}$ is the matrix of loading weights.

The PLS regression method as described here can be extended to handle several $y$-variables simultaneously (PLS2). The methods are very similar, the only modification is that instead of maximising the covariance between $y$ and linear functions of $\mathbf{x}$, one now needs to optimise the covariance between two linear functions, one in $\mathbf{x}$ and one in $\mathbf{y}$. For interpretation purposes, this extension may be advantageous, but for prediction purposes, it is usually better to calibrate for each $y$-variable separately. The focus in this book will be on the single-$y$ situations and the PLS2 method will not be pursued further here.

It should be mentioned that a slightly modified PLS algorithm involving a so-called inner relation be-

# Data compression by PCR and PLS

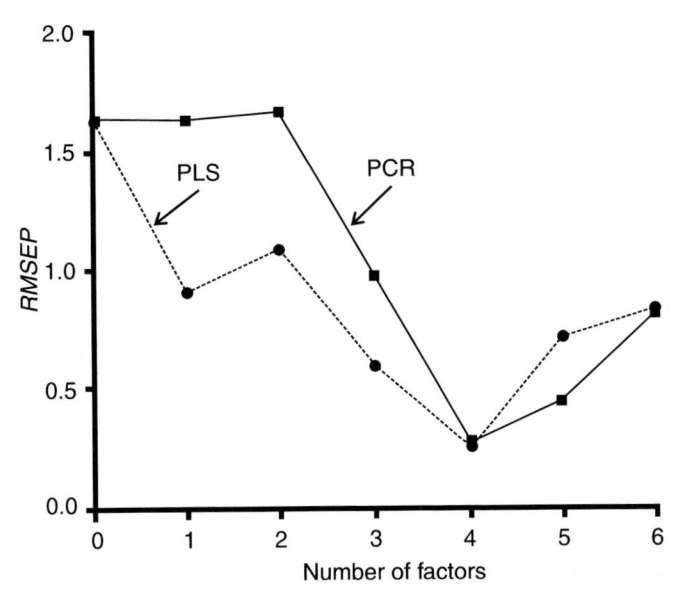

**Figure 5.2.** Prediction ability of PCR and PLS for $A = 0$, 1, ..., 6. The *RMSEP* is the root mean square error of prediction (see Chapter 13), and measures the prediction ability. Small values of *RMSEP* are to be preferred. Note that the *RMSEP* follows a similar curve to the prediction error in the conceptual illustration in Figure 4.3.

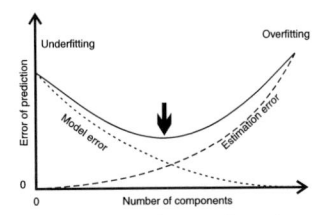

**Figure 4.3.** The estimation error (– – –) and model error (· · ·) contribute to the prediction error (——). If too large a model is used, an overfitted solution is the result. The opposite is called underfitting. Usually a medium size model is to be preferred.

tween **X**-scores and **y**-scores has been developed [see Esbensen *et al.* (2000) and the publications in the reference list by S. Wold]. The predictions obtained by this algorithm are identical to predictions obtained by the method described above. For PLS2 calibrations, this other algorithm provides an extra tool, the inner relation plot, which can be useful for detecting outliers and non-linearities (see Chapter 14).

## 5.5 Comparison of PCR and PLS regression

Experience with NIR data has shown that PLS regression can give good prediction results with fewer components than PCR. A consequence of this is that the number of components needed for interpreting the information in **X** which is related to **y** is smaller for PLS than for PCR. This may in some cases lead to simpler interpretation. Using the optimal number of components in each case, however, the two methods often give comparable prediction results. Several simulation studies support this conclusion. Cases exist in practice where PLS regression is better as do cases where PCR is better, but the difference is not usually large.

A very typical situation is illustrated in Figure 5.2. Here, PLS and PCR are compared on a small NIR data set of wheat with 12 calibration samples, 26 test samples and six NIR wavelengths [the data are taken from Fearn (1983)]. The focus is on predicting protein percentage. The condition number (*CN*) is equal to 131 and the variance inflation factors (*VIF*s) for the six *x*-variables are 1156, 565, 1066, 576, 16 and 83. These values are all very high showing that the data are highly collinear. Both methods start out with poor prediction results for one component (*RMSEP* = 1.6 for PCR and *RMSEP* = 0.9 for PLS) (*RMSEP* is the root mean square error of prediction (see Chapter 13). The methods improve as we introduce more components, PLS a bit faster than PCR, before both methods end up at about the same optimal prediction ability for $A = 4$ (*RMSEP* = 0.28). Then the PLS regression errors go more steeply up towards the same solution as PCR for $A = 6$ (*RMSEP* is close to 0.8). Note that both methods are identical to MLR for the maximal number of components ($A = 6$).

From a computational point of view, PLS is usually faster than PCR. For large data sets this aspect may be of some value.

From a theoretical point of view, PCR is better understood than PLS. Some recent theoretical aspects of PLS and its relation to PCR are discussed in Næs and Helland (1993), Høskuldsson (1996), Helland

and Almøy (1994), Almøy (1996), Helland (1988), Manne (1987), Lorber *et al.* (1987) and de Jong (1995). Comparisons with other methods such as ridge regression (RR) can be found in Frank and Friedman (1993) and Næs *et al.* (1986).

## 5.6 Continuum regression

In 1990, Stone and Brooks proposed a new data compression method for calibration, which they termed continuum regression (CR). Essentially, this method, or rather class of methods, is based on the same model structure as above for PLS/PCR (Formula 5.1 and Figure 5.1). The only difference is a more flexible criterion for computing components. The components are computed sequentially as was described for the PLS method above, with orthogonality constraints producing different components at each step.

Instead of defining a single criterion to be optimised, Stone and Brooks proposed a class of criteria indexed by a parameter $\alpha$ varying between 0 and 1. The CR criterion which is to be maximised is

$$\text{cov}(\mathbf{w}^t\mathbf{x}, y)^2 \times \text{var}(\mathbf{w}^t\mathbf{x})^{[\alpha/(1-\alpha)]-1} \quad (5.7)$$

where maximisation is done over all possible directions $\mathbf{w}$ with length equal to one. As above, $\mathbf{x}$ and $y$ represent centred data. The cov and var functions here refer to the empirical covariance and variance functions, respectively (see Appendix A).

For $\alpha$ equal to 0, maximising the criterion (5.7) is identical to maximising the correlation between $y$ and linear functions of $\mathbf{x}$. As described in Appendix A, this gives the MLR solution (or LS solution). In this case, only one component $\mathbf{w}$ can be extracted and this is proportional to the LS regression coefficient vector $\hat{\mathbf{b}}$.

When $\alpha = 1/2$ the criterion corresponds to maximising the covariance. This is exactly how PLS was defined. When $\alpha$ approaches 1, the criterion becomes more and more dominated by the variance component of the product in (5.7), and PCR is obtained.

Thus, continuum regression is a unified framework for all the methods PCR, PLS and MLR. From a conceptual point of view CR is then very attractive. In practice, one needs to determine the best possible choice of $\alpha$. This can, as was suggested by Stone and Brooks (1990), be done by cross-validation or by prediction testing. One simply computes the prediction ability of the method for a number of different choices of $\alpha$ and tries to find the best possible choice. Note that for continuum regression, one then needs to cross-validate two parameters, the number of components, $A$, and the parameter $\alpha$.

Some practical results of CR are presented in Stone and Brooks (1990). The method, however, has so far been little used in applications in the literature.

# 6 Interpreting PCR and PLS solutions

## 6.1 Loadings and scores in principal component analysis (PCA)

Suppose we have observations on $K$ spectral variables for each of $N$ samples, with $x_{ik}$ being the value of the $k$th variable for the $i$th sample. Then the $N \times K$ data matrix $\mathbf{X}$ has the spectral data for the $i$th sample as its $i$th row. Principal component analysis, the data compression step in PCR, seeks to model the $N$ observed spectra as linear combinations of a much smaller number, $A$, of so-called components. The centred data matrix $\mathbf{X}$ is decomposed as

$$\mathbf{X} = \hat{\mathbf{T}}\hat{\mathbf{P}}^t + \hat{\mathbf{E}} = \hat{\mathbf{t}}_1\hat{\mathbf{p}}_1^t + \hat{\mathbf{t}}_2\hat{\mathbf{p}}_2^t + \ldots + \hat{\mathbf{t}}_A\hat{\mathbf{p}}_A^t + \hat{\mathbf{E}} \quad (6.1)$$

where the $N \times 1$ column vector $\hat{\mathbf{t}}_a$ contains the scores of the $N$ samples for the $a$th component, the $K \times 1$ vector $\hat{\mathbf{p}}_a$ contains the loading values for the $a$th component and the error matrix $\hat{\mathbf{E}}$ is the unexplained part of $\mathbf{X}$. This residual matrix corresponds to the dimensions in $x$-space which are not accounted for by the first $A$ components. The components in $\hat{\mathbf{T}}$ can be thought of as "underlying" or "latent" variables, which can be multiplied by $\hat{\mathbf{P}}$ in order to approximate the systematic variation in $\mathbf{X}$.

One interpretation follows directly from this. Each of the samples (rows of $\mathbf{X}$) is being represented as a linear combination of the vectors $\hat{\mathbf{p}}_a$, $a = 1, \ldots, A$, just as though the sample was a mixture of $A$ constituents and $\hat{\mathbf{p}}_a$ was the spectrum of the $a$th constituent. Thus it is possible to interpret the $\hat{\mathbf{p}}_a$ as the "spectra" of the factors/components/latent variables that model $\mathbf{X}$. There is, however, a catch to this. Even when the samples truly are mixtures following an additive

**AA**
LWR: locally weighted regression
NIR: near infrared
PCA: principal component analysis
PCR: principal component regression
PLS: partial least squares
*RMSEP:* root mean square error of prediction

It is much easier to discuss interpretation with a particular context in mind, and this chapter is written as though the *x*-variables must be spectra. The same ideas are applicable far more generally, however.

Beer's law, there is nothing that makes an individual $\hat{\mathbf{p}}$ correspond to an individual physical constituent. The problem is that it is hard to establish the basic ingredients of a set of mixtures just from observing the mixtures, and impossible if you allow negative proportions as the mathematics of PCA does. Any mixture of A, B and C could equally well be expressed as a mixture of A + B, B + C and C + A. Thus you will see in the loadings the spectral features of the constituents, but features from all of them may well be present in any one $\hat{\mathbf{p}}$.

The $\hat{\mathbf{p}}_a$s are orthogonal ($\hat{\mathbf{p}}_a' \hat{\mathbf{p}}_b = 0$ for $a \neq b$) and normally scaled to have length 1, so that $\hat{\mathbf{p}}_a' \hat{\mathbf{p}}_a = 1$. As was described in section 5.3, each $\hat{\mathbf{t}}_a$ can be found by multiplying $\mathbf{X}$ by $\hat{\mathbf{p}}_a$, i.e.

$$\hat{\mathbf{t}}_a = \mathbf{X}\hat{\mathbf{p}}_a \tag{6.2}$$

which means that the way to find the score of the *i*th sample on the *a*th factor is to combine the *K* original spectral measurements on the sample using the *K* elements of $\hat{\mathbf{p}}_a$ as weights. This gives a second way of interpreting $\hat{\mathbf{p}}_a$, as weights that define the contribution of each wavelength/variable to the constructed component.

For interpretation, information from the score matrix $\hat{\mathbf{T}}$ is usually presented in two-dimensional scatter-plots of one component versus another. For instance, one plot corresponds to principal component 1 versus component 2, while another one corresponds to component 3 versus component 4 etc. An example of a scores plot can be found in Figure 6.1. Often, a very high percentage of the relevant spectral variability (as measured by the sum of the corresponding eigenvalues) is gathered in the first few principal components so only a few of these plots are necessary. The scores plots are frequently used for detecting such aspects as similarities, differences and other interesting relationships among the samples. This is much

6.1 Loadings and scores in principal component analysis (PCA)

# Interpreting PCR and PLS solutions

(a) Starch 100%

Water 100%  Fish meal 100%

(b)

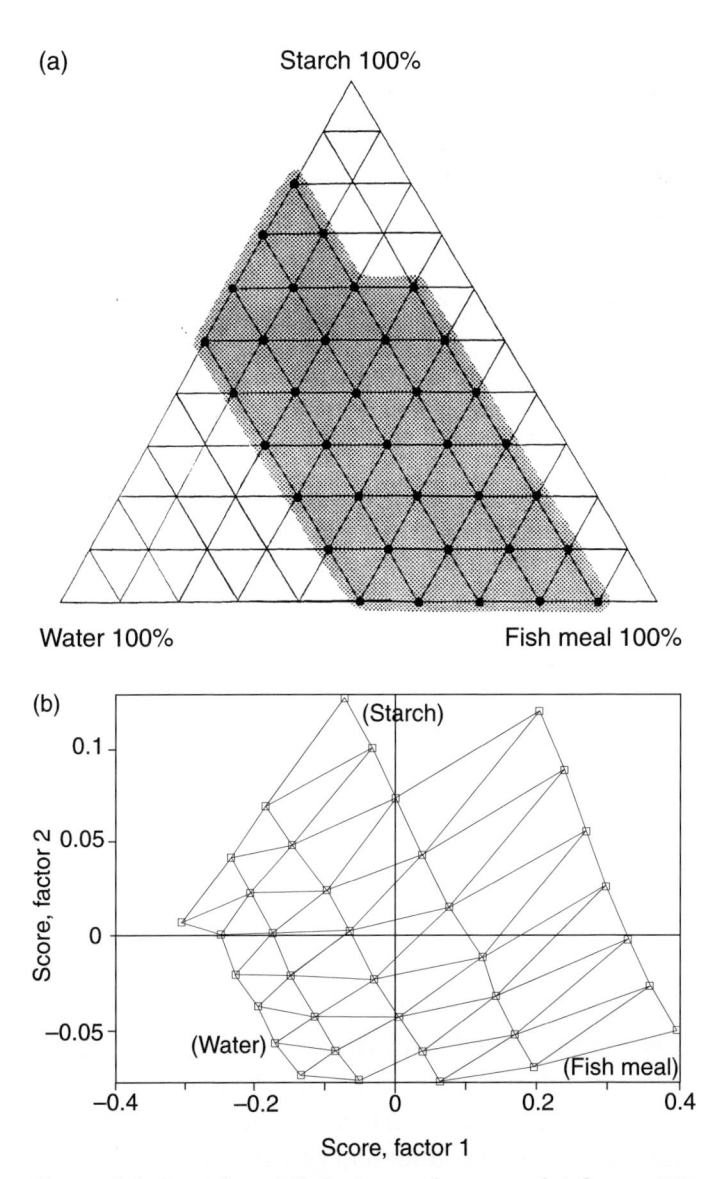

Figure 6.1. Experimental design and scores plot for an NIR data set. (a) The shaded region shows the experimental region used for NIR analysis. Three ingredients, fish protein, starch and water were mixed in different proportions. (b) The PCA scores plot of the data in (a). The data were scatter-corrected (see Chapter 10) prior to PCA.

**6.1 Loadings and scores in principal component analysis (PCA)**

easier in a compressed data space than in the original *K*-dimensional space. Outliers, subgroups/clusters etc. can sometimes also be detected using such plots. Outlier detection is covered in Chapter 14 and clustering in Chapter 18.

The loadings $\hat{\mathbf{P}}$ are also often presented in two-dimensional scatter-plots of pairs of components. An alternative that is more often used in spectroscopy is to plot each loading vector versus wavelength number (see Section 6.4). In this way the spectral pattern of each component is easier to see. The loadings are used for interpreting the relationship between the scores and the original variables. They can, for instance, be used to tell which of the variables are most relevant for the different components in the data set [see formula (6.2)]. If the focus is to find clusters in the data set, loading plots can be useful for identifying which variables characterise the different clusters.

An example illustrating the use of PCR for interpretation in spectroscopy is given in Section 6.4.

## 6.2 Loadings and loading weights in PLS

What is surprising about PCA is that choosing orthogonal scores also gives us orthogonal loadings, that is $\hat{\mathbf{p}}_a^t \hat{\mathbf{p}}_b = 0$ for $a \neq b$. This double orthogonality only happens with PCA, when the $\hat{\mathbf{p}}_a$s are determined as eigenvectors of $\mathbf{X}^t\mathbf{X}$. With PLS we have to choose between orthogonal loadings or uncorrelated scores, and it is usually the latter that gets the vote. This is the algorithm described above. The other algorithm can be found in, for instance, Martens and Næs (1989). The two algorithms give equivalent prediction results, although they provide slightly different tools for interpretation.

The second set of PLS loadings, the loading weights $\hat{\mathbf{w}}_a$ are used to construct the scores, and so have an interpretation similar to the second one given above for PCA loadings. However, it is not quite as simple as this. The vector of weights $\hat{\mathbf{w}}_1$ is computed from $\mathbf{X}$ directly and applied to the original spectral measurements to construct the scores on PLS component 1. The weights in $\hat{\mathbf{w}}_2$ and corresponding scores are, however, not computed from $\mathbf{X}$, but from the residuals $\mathbf{X} - \hat{\mathbf{t}}_1 \hat{\mathbf{p}}'_1$. For the rest of the components, the loading weights are computed using other residual matrices. This aspect makes it quite complicated to compare and then interpret the *loading weights*. The loadings $\hat{\mathbf{P}}$ on the other hand can be found by regressing $\mathbf{X}$ directly onto the final $\hat{\mathbf{T}}$ matrix. For interpretation, our preference is therefore to use the loadings, $\hat{\mathbf{P}}$, rather than using the loading weights $\hat{\mathbf{W}}$. Another advantage of using $\hat{\mathbf{P}}$ is that it is directly comparable to $\hat{\mathbf{q}}$. The reason for this is that $\hat{\mathbf{q}}$ can be obtained in the same way by regressing the original $\mathbf{y}$ onto the common scores matrix $\hat{\mathbf{T}}$.

> We recommend the use of loadings for interpretation in PLS.

Much of the time the loadings and loading weights will be similar to each other, and indeed similar to the loadings that would have been obtained from PCA. When they differ it shows that PLS is choosing different factors to PCA. The spectral information related to $\mathbf{y}$ is in such cases not at the most variable wavelengths in the spectrum. PCA only tries to model $\mathbf{X}$, and large loading values correspond to spectral regions with large variability. PLS compromises between modelling variance in $\mathbf{X}$ and correlation between $\mathbf{y}$ and $\mathbf{X}$, thus big loadings ($\hat{\mathbf{p}}$ or $\hat{\mathbf{w}}$) may correspond to high variance, high correlation or both. Regions of discrepancy between $\hat{\mathbf{p}}$ and $\hat{\mathbf{w}}$ show where PLS is heavily influenced by the $\mathbf{X}$–$\mathbf{y}$ correlation.

## 6.3 What kind of effects do the components correspond to?

In many practical situations, it can be observed that the number of components needed to obtain good prediction results is much larger than the number of chemical components in the sample. What do these extra components correspond to?

If Beer's law applies and all components absorb light in the region concerned, the number of underlying components should correspond to the number of varying chemical components minus one (because the concentrations of all constituents sum up to 100%). If more components are needed, this must be due to other effects. An important and frequently occurring effect in spectroscopy is the light scatter effect due to differences in particle size, sample thickness etc. (see Appendix B). This kind of effect can be very strong in applications of, for instance, diffuse reflectance NIR. Other possibilities that may also lead to more components are interactions among the chemical constituents and non-linear relations between the chemical concentrations and the spectral measurements.

Non-linearity between chemistry and spectral readings is illustrated in Figure 6.1. This example is taken from a designed NIR experiment with mixtures of three chemical components, fish protein, starch and water. If Beer's law applied, a principal component analysis of the NIR data would then result in a two-dimensional plot with the same shape as the design of the experiment. Comparing Figures 6.1(a) and 6.1(b) it is clear that this is not exactly the case here, but it is also clear from the figure that the first two principal components contain very much of the information in the full design. The difference between the two plots is a non-linear effect; the contour lines in the PCA scores plot are no longer equidistant and they are clearly curved.

Using PCR on these data with protein content as $y$, the best prediction results (10 components) for an independent test set gave $RMSEP = 1.00$. Using only two components in the regression gave an $RMSEP$ of about 3.00. In other words, even though the main in-

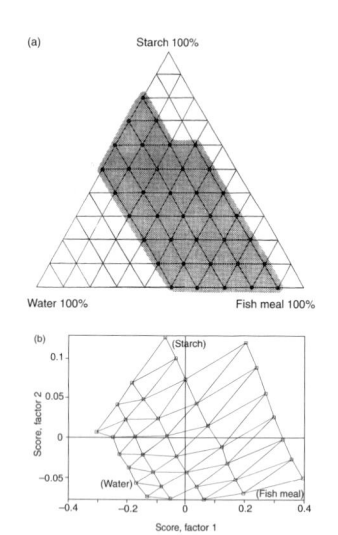

**Figure 6.1.** Experimental design and score plot for an NIR data set.

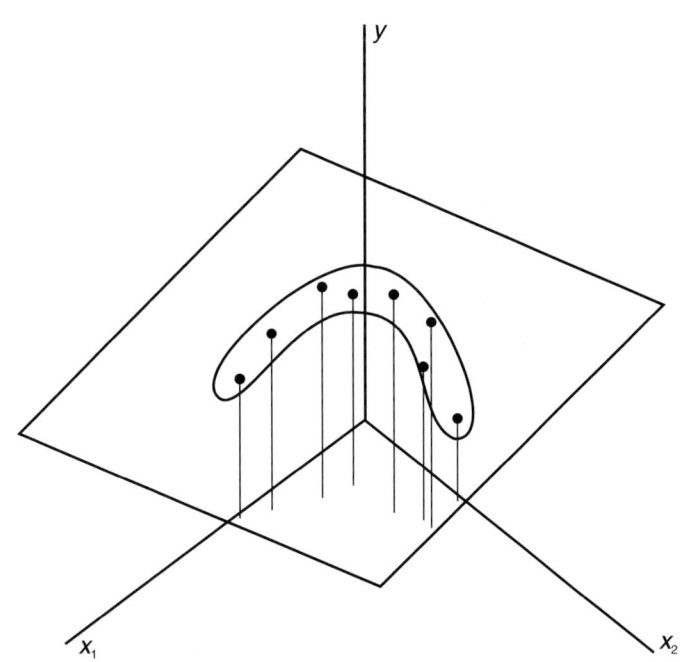

Figure 6.2. In this example, the two x-variables each have a non-linear relationship with y. Since all observations lie in a plane, the y variable is, however, linearly related to $x_1$ and $x_2$ when the two x-variables are considered together in a multivariate linear model.

formation seems to be in the first two dimensions, we need more components to obtain good results. The first two components were then used in a non-linear calibration, using the locally weighted regression (LWR) method (see Chapter 11) and in that case we obtained $RMSEP = 0.67$. This represents a 33% improvement as compared to the best possible PCR using a much larger number of components. This proves what the comparison of figures indicates, namely that the first two components contain the most relevant information, but in a non-linear way. It therefore needs a non-linear treatment in order to give the best possible results. The results also show that more principal components in the model can sometimes compensate for some of the non-linearities seen in the first

**6.3 What kind of effects do the components correspond to?**

few components. This is one of the reasons why linear regression methods such as PCR and PLS give such good results in remarkably many practical situations.

This latter phenomenon is also illustrated conceptually in Figure 6.2 [see also Martens and Næs (1989)]. In this illustration one can see that $y$ is non-linearly related to both $x_1$ and $x_2$. These are called univariate non-linearities. The first principal component is also non-linearly related to $y$. When both components are used, it is, however, possible to find a linear predictor which fits perfectly to $y$. This also shows that even if there are strong univariate non-linearities, the multivariate relation may be linear. In some cases, the multivariate relation is so non-linear that a non-linear method is needed to model it successfully.

It should also be mentioned that there is an important connection between overfitting and the appearance of the one-dimensional loading plots often used in spectroscopy [see Martens and Næs (1987)]. For the components corresponding to the largest eigenvalues, the shapes of these plots will look like smooth spectra, but as one goes beyond the optimal number of components, the noise starts to become clearly visible. Therefore, it is a sensible practice to let component selection based on regular validation criteria (see Chapter 13) be accompanied by visual inspection of the loadings. This is particularly true in spectroscopy [see, for instance, Martens and Næs (1987)].

## 6.4 Interpreting the results from a NIR calibration

The present section presents an example of how a PCA/PCR analysis of NIR data can be interpreted. The example is constructed from a designed experiment of ingredients with known spectral shapes. This makes it possible to check the interpretation of the multivariate analysis against information available prior to the analysis. This is not always the case in a practical situation.

An important reason for wanting to interpret a calibration is to make sure that the results obtained are reasonable compared to information available from theory or prior experience. This type of interpretation can reduce chances of making obvious mistakes and plays an important role when validating a model. In some cases, however, unexpected and new ideas and results can be obtained from an interpretation of a calibration model.

Figure 6.3 shows the design of the NIR experiment. It is based on mixtures of casein, lactate and glucose. Each sample in the design was analysed by a NIR scanning instrument: a Technicon InfraAlyzer 500 operating in the range between 1100 and 2500 nm. The NIR spectra of pure lactate, casein and glucose are shown in Figure 6.4.

All the 231 spectra were subjected to principal component analysis after scatter correction using the multiplicative scatter correction (MSC, Chapter 10) technique. The score plot of the first two components is shown in Figure 6.5. The score plot is quite regular and similar to the design in Figure 6.3. Especially the region close to glucose and lactate looks like the corresponding part of the design triangle. The first two components describe 76.5% and 23.1% of the total variation, respectively (sum = 99.6%). Knowing from theory that if Beer's law applies, one would obtain an exact triangular representation with straight contour lines in the first two components, this result means that scatter-corrected data are quite well modelled by Beer's law. The same computation on non-corrected data gave quite similar results showing that the scatter correction had a relatively small effect in this case.

Thus, the scatter-corrected and mean-centred spectral matrix $\mathbf{X}$ can be approximated by the sum of

**6.4 Interpreting the results from a NIR calibration**

Casein 100%

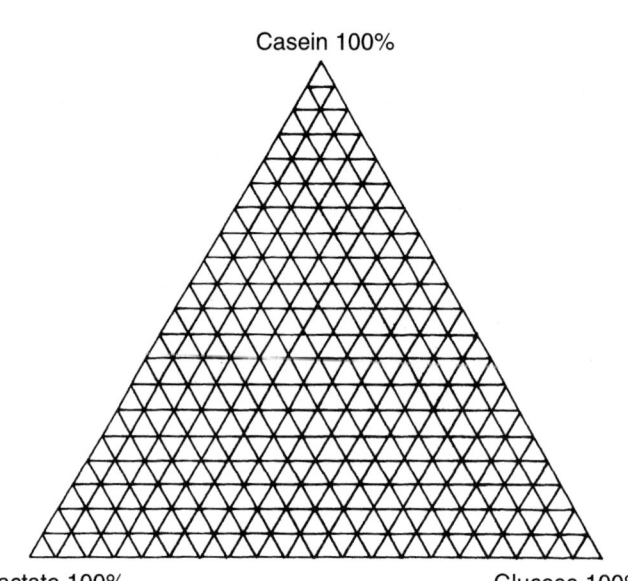

Lactate 100%  Glucose 100%

**Figure 6.3.** The design of a NIR experiment based on mixtures of casein, lactate and glucose. Each vertex corresponds to an experimental point.

the loading vector in Figure 6.6(a) multiplied by the first principal component score vector and the loading in Figure 6.6(b) multiplied by the second principal component, i.e. $\mathbf{X} \approx \hat{\mathbf{t}}_1 \hat{\mathbf{p}}_1' + \hat{\mathbf{t}}_2 \hat{\mathbf{p}}_2'$. Non-linearities seem to play a relatively minor role in this application.

From Figure 6.5 it is clear that the first component is close to being a pure glucose component, while the second component is related to the difference between lactate and casein. This means that the first loading vector should be very similar to the spectrum of glucose and that the second loading vector should be similar to the difference between the spectra of lactate and casein. Comparing Figure 6.6 with Figure 6.7 both these conclusions are clearly verified.

The $y$-loadings $\hat{q}_1$ and $\hat{q}_2$ correspond to scores $\hat{\mathbf{t}}_1$ and $\hat{\mathbf{t}}_2$ which account for quite different proportions of the total variance. To make it possible to compare $y$-loadings for the different components, it may there-

# Interpreting PCR and PLS solutions

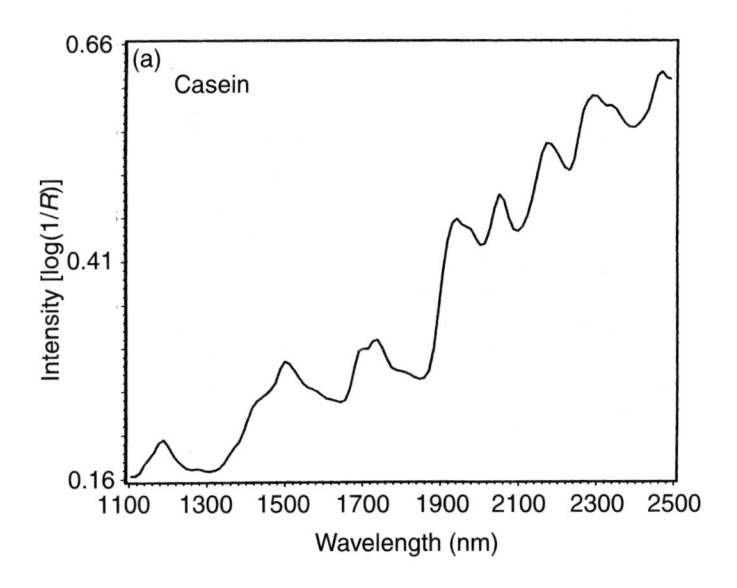

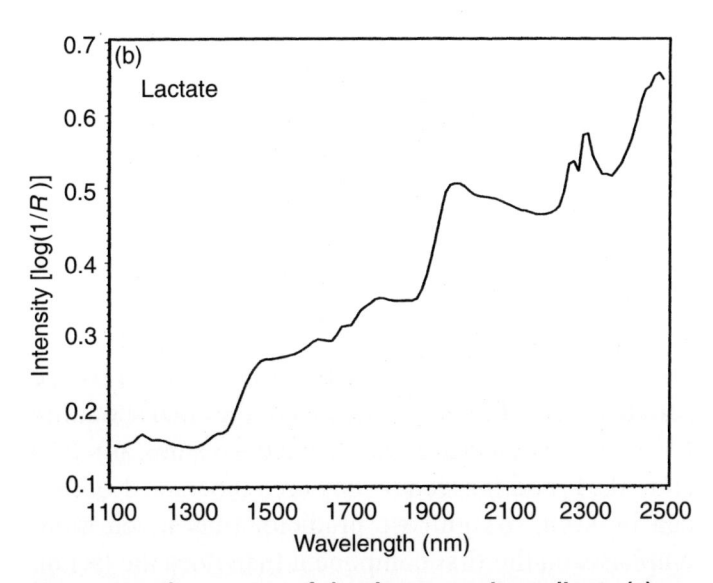

Figure 6.4. The spectra of the three pure ingredients (a) casein, (b) lactate and (c) glucose (see over) in the experiment. Casein gives, as a protein, a complex spectrum with many bands originating from many functional groups (e.g. C–H, O–H, N–H, C=O etc.). The lactate spectrum is dominated by one band at 1950 nm.

## 6.4 Interpreting the results from a NIR calibration

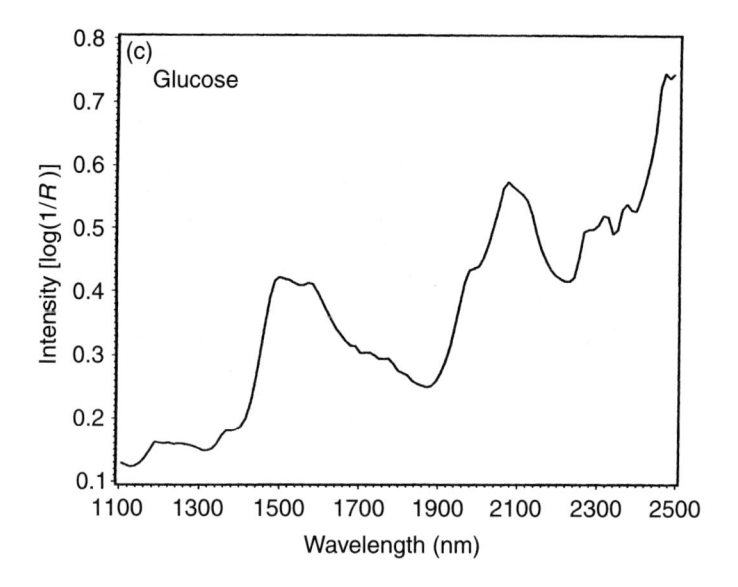

**Figure 6.4(c).** The spectra of the third pure ingredient (c) glucose in the experiment. The glucose spectrum is dominated by two bands or peaks, one at 1500–1600 nm (O–H stretching, first overtone and O–H stretching first overtone, intermolecular hydrogen bonds) and another at 2100 nm (O–H stretching + deformation, combination band).

fore be advantageous to multiply the $\hat{q}$-values by the square root of the sum of squares of the corresponding $\hat{t}$-values. For glucose, these scaled $\hat{q}$-values are 2.10 and –0.31, and for lactate they are –0.65 and 1.65. As can be seen, the glucose predictor puts much more emphasis on the first component than does the lactate predictor. This is again natural from the plot in Figure 6.5, where it was shown that the first component is very close to being a pure glucose component. For lactate, the second component was most important, which was also to be expected from the score plot. The first two components describe almost 99% of the variance of $y$, but an even better result was obtained

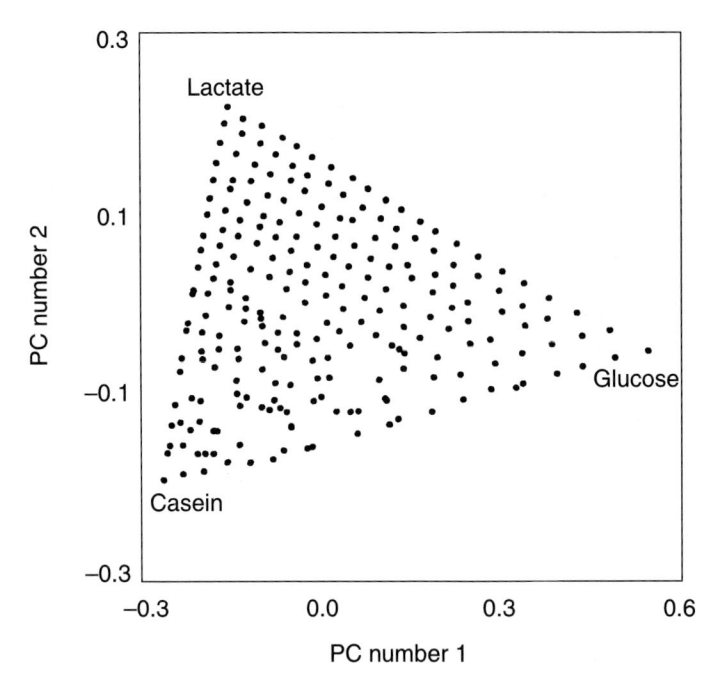

**Figure 6.5.** PCA score plot of the NIR spectra of the samples in the design in Figure 6.3. The data were scatter-corrected prior to PCA.

by including two more components in the PCR equation. This shows that even in this simple case, the linearity is not perfect and can be improved upon by using more components.

To sum up, this example has first of all shown that the NIR data matrix in this case can be quite adequately modelled by a linear Beer's law. It has also shown that glucose is closely related to the first component and thus responsible for more spectral variability than the other two constituents. It has also indicated which wavelengths are affected by the different constituents. These results were verified by comparing with prior information about the design and spectral patterns of the constituents. In addition, the score plot reveals no outliers and no strange or abnormal

**6.4 Interpreting the results from a NIR calibration**

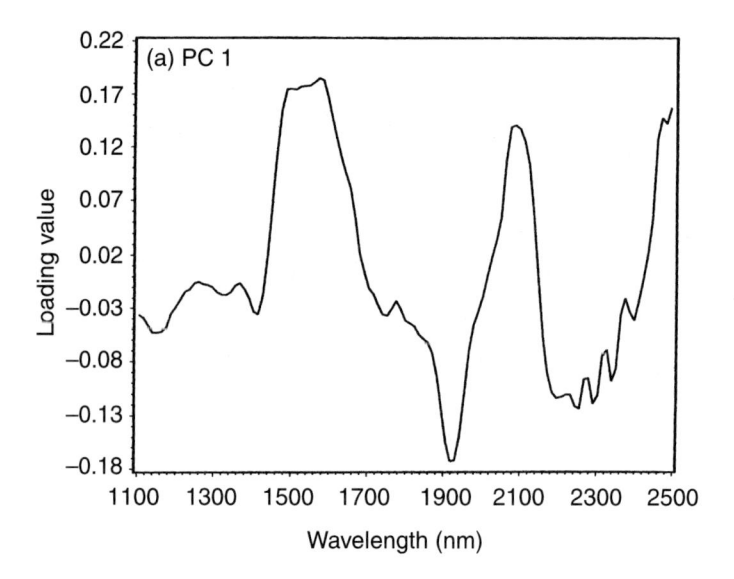

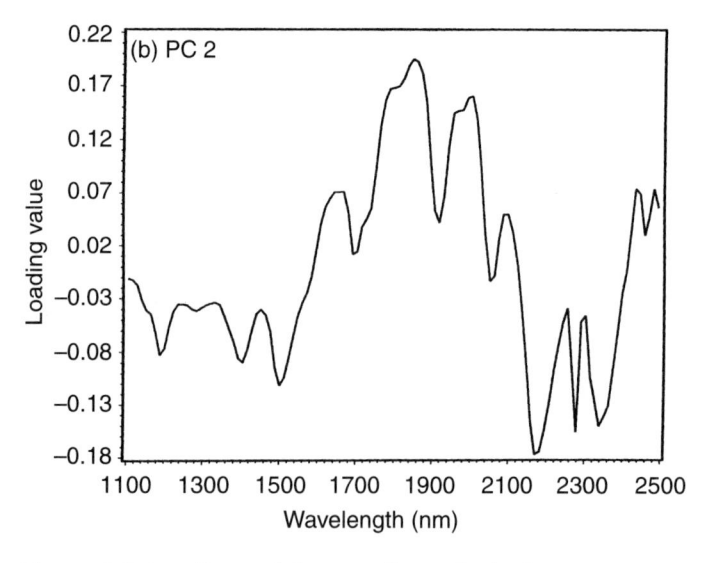

**Figure 6.6.** Loadings of the two first principal components. (a) presents the first loading vector and (b) the second loading vector.

structures, strengthening our confidence in the results. The interpretation of scores, **X**-loadings and **y**-loadings also fit well together.

# Interpreting PCR and PLS solutions

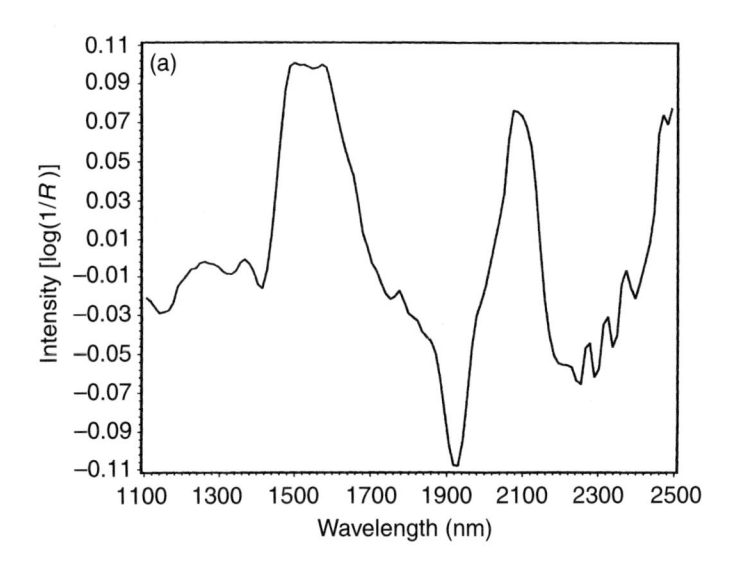

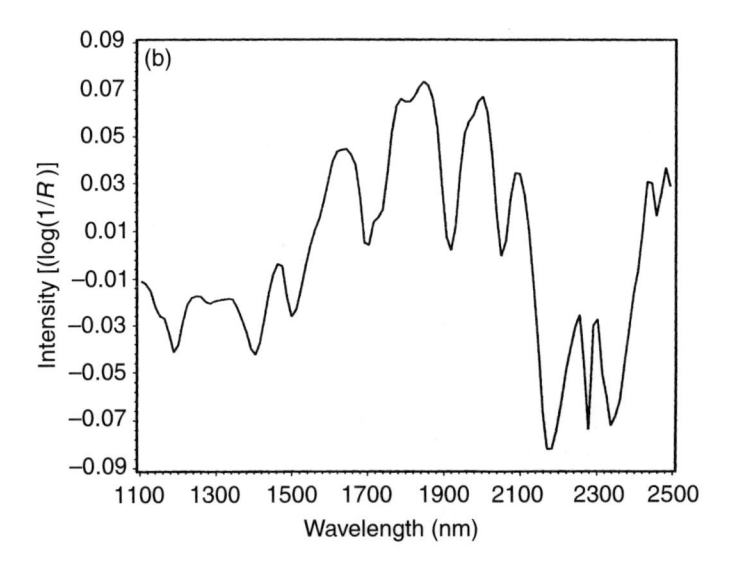

**Figure 6.7.** (a) The mean-centred glucose spectrum. The glucose spectrum minus the average spectrum is dominated by the two positive bands (1500–1600 nm and 2100 nm) from glucose and the negative band from lactate (1950 nm). (b) The difference between the spectra of lactate and casein. This is characterised by the many absorbance bands from the complex casein spectrum.

## 6.4 Interpreting the results from a NIR calibration

For further examples of interpretation of plots from NIR spectroscopy, we refer to, for example, Miller *et al.* (1993), Beebe *et al.* (1998), Osborne *et al.* (1993) and Williams and Norris (1987).

# 7 Data compression by variable selection

## 7.1 Variable selection for multiple regression

The subject of this chapter is selection of variables for use in regular MLR calibration. When the *x*-data are spectra, variable selection means using data from a small number of wavelengths, and is often referred to as wavelength selection. This is an alternative strategy to applying PCR or PLS to all variables (see Figure 4.2). The variable selection methods try to find the most relevant and important variables and base the whole calibration on these variables only.

One situation in which variable selection is clearly the preferred approach is when we can reduce the cost of future measurements by observing only the selected variables. An example would be when the calibration data come from a scanning spectrophotometer, but the aim is to use the calibration on several cheaper filter instruments with a limited number of wavelengths. In such cases, selecting the most powerful combination of wavelengths is very important.

The discussion below deals with the statistical and computational aspects of variable selection methods. One should, of course, not lose sight of the chemistry amongst all this: equations that use interpretable variables are distinctly preferable to those that do not, however clever the algorithm that found the latter. There are two main problems to discuss, which criterion to use for optimisation of the number of the variables and which strategy to use for the search. The two are not totally independent, as will be seen below.

*AIC:* Akaike information criterion
*LS:* least squares
*MLR:* multiple linear regression
*MSEP:* mean square error of prediction
*PCR:* principal component regression
*PLS:* partial least squares
*RMSECV:* root mean square error of cross-validation
*RMSEP:* root mean square error of prediction
*RSS:* residual sum of squares
*SMLR:* stepwise multiple linear regression

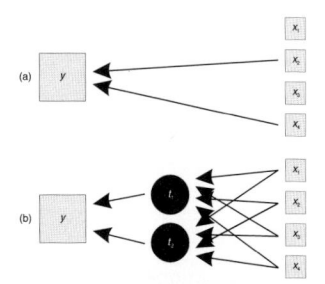

Figure 4.2. Conceptual illustration of the difference between methods for solving the multicollinearity problem.

The basic model underlying the methodology presented in this chapter is the standard linear model given by

$$y = b_0 + b_1 x_1 + b_2 x_2 \ldots + b_K x_K + f \qquad (7.1)$$

We assume that there are $N$ observations of the vector $(y, \mathbf{x}^t)^t$ available for estimation of the regression coefficients.

The problem is to find a subset of the $x$s that can be used to predict $y$ as precisely and reliably as possible in situations with many collinear $x$-variables. For most of this chapter, estimation of regression coefficients will be done by the regular least squares criterion [LS, see Appendix A and Weisberg (1985)]. This criterion seeks coefficients that minimise the sum of squared differences between the observed and fitted $y$s. At the end of the chapter we will discuss how variable selection also can be combined with PLS/PCR for model estimation. The latter is a new and interesting development, which deserves more attention in the literature.

## 7.2 Which criterion should be used for comparing models?

### 7.2.1 Using residual sum of squares (RSS) or $R^2$

To compare two equations with the same number, $p$, of variables, it is generally reasonable to compare the residual sums of squares (*RSS*, Appendix A)—the smaller the value, the better the fit. It might also seem reasonable to compare equations with different numbers of variables in this way, but unfortunately it is not. The best equation with $p + 1$ variables will inevitably have an *RSS* at least as small as, and usually smaller than, the best one with $p$. Thus using *RSS* (or equivalently $R^2$, see Appendix A) as the criterion to compare equations would generally result in a model with too many variables in it. This would lead to equations that fit the data, but would not necessarily lead to improved predictions (overfitting, see Section 4.3).

Using the residual variance estimate $\hat{\sigma}^2 = RSS/(N-p-1)$ does not help much either, even though both the numerator $RSS$ and the divisor $N-p-1$ decrease as $p$ increases. In fact what typically happens to the residual variance is that it decreases for a while as we add terms to the equation and then stabilises, staying roughly constant or declining very slowly as further terms are added. The intuitive explanation for this is that $\hat{\sigma}^2$ estimates the variance of the error term in the model equation, not its prediction ability (see Appendix A).

### 7.2.2 Alternative criteria to RSS based on penalising the number of variables

The problem can be solved in (at least) two ways. Either we can find a criterion that penalises additional variables in some appropriate way, or we can use a validation procedure, involving either cross-validation or a separate test set, to choose between equations with different numbers of variables (see also Chapter 13).

A simple way to penalise additional variables is to use $C_p$, which is defined by

$$C_p = RSS/\hat{\sigma}^2 + 2p - N \qquad (7.2)$$

Here $\hat{\sigma}^2$ is an estimate of the "correct" residual variance of model (7.1), $p$ is the number of variables in the subset equation and $RSS$ is the corresponding residual sum of squares. This criterion is known as Mallows $C_p$ and is an estimate of the average mean square error of prediction (*MSEP*, see Chapter 13) divided by the residual error variance in model (7.1).

**7.2** Which criterion should be used for comparing models?

The criterion $C_p$ is a special case of the Akaike Information Criterion [*AIC*, Akaike (1974)], which may be used for choosing between statistical models in a more general setting.

The value of $C_p$ can be computed and used to compare models of different size. Good models will give small values of $C_p$. Mallows (1973) has suggested that good models have a $C_p$ value close to $p$.

The obvious difficulty in applying the criterion is the need for a value for $\hat{\sigma}^2$. The usual textbook recommendation is to use the residual variance $RSS/(N - K - 1)$ from the full equation, i.e. the one using all $K$ $x$-variables. This is, however, not possible when the number of samples is smaller than the number of variables. One solution is to plot residual variance against number of terms in the equation. As noted above, this should decline, then stabilise. One can then use the level at which it stabilises as $\hat{\sigma}^2$.

F-tests (see Appendix A.4) are also often used for comparing models of different size if one of the models is a sub-model of the other one. One simply computes the F-statistic for the terms not present in the smaller model and uses this as the criterion. If F is not significant, one concludes that these extra terms are not useful. Especially in connection with forward or backward strategies for selecting variables, F-tests are important. An advantage of using the F-test over comparing $C_p$ values is that it provides information about the statistical significance of the extra terms added to the models. F-tests can, however, not be used for comparing models of the same size and models which are not nested. There is a close connection between $C_p$ and the F-statistic, which is described in, for instance, Weisberg (1985).

### 7.2.3 Criteria based on a test set or cross-validation

The criteria to be discussed here are based on computations made on a separate set of data (prediction testing) or a procedure "simulating" prediction testing using the calibration data themselves. This latter strategy is called cross-validation and is based on eliminating samples successively for testing purposes. We refer to Chapter 13 for a more thorough discussion of these concepts.

A suggested procedure for using a separate test set in this situation is the following. Set aside a number of samples to make up the test set. Then use the calibration set to determine the variables that give the best equations with 1, 2, ..., $p$ terms (using, for instance, *RSS*). Now compare this sequence of equations by using the square root of the average sum of squared prediction errors for the test set (*RMSEP*, see Chapter 13). For the smaller models, the *RMSEP* will typically decrease as we add variables, then it will stabilise around the optimal value before it increases as we start overfitting (see Figure 4.3). A reasonable model choice will be the one that minimises *RMSEP*. If a smaller model has a similar *RMSEP* value, this will usually be an even better choice.

The alternative is to use cross-validation (see Chapter 13). This is a technique closely related to prediction testing using the calibration data only. The procedure goes as follows. Delete one sample from the set, use the remaining $N-1$ samples to estimate the coefficients of the equation, and predict the deleted sample. Repeat the same procedure for each sample and sum the squared prediction errors. The square root of the average sum of squares of the differences between measured and predicted values is called *RMSECV*. Like *RMSEP*, *RMSECV* will typically pass through a minimum as the number of variables is increased.

A variant of this procedure that it would in principle be possible to use is to apply *RMSEP* or *RMSECV* (rather than *RSS*), also to choose between competing equations with the same number of terms.

**7.2 Which criterion should be used for comparing models?**

This seems undesirable on two grounds, one theoretical—the more the test set is used, the more over optimistic is any estimate of performance based on *RMSEP*—and one practical—it would increase the computational cost considerably. A remedy sometimes used for solving the former problem, but which is quite data intensive is to use a separate set for testing the performance only [see Nørgaard and Bro (1999)]. This would require three sets of data, one for fitting, one for selecting the model and one for validating the performance. Alternatively, the calibration set may be used for both the two first aspects, i.e. fitting and model selection (by cross-validation). Then only a second set is needed for testing.

## 7.3 Search strategies

### 7.3.1 Best subset selection

Best subset selection is a strategy based on comparing all possible models (up to a certain size $p$) with each other using one of the criteria discussed above. As was mentioned, $R^2$ and *RSS* are useful for comparing models of equal size, but not for models of different size. $C_p$ is therefore usually to be preferred since it penalises the number of parameters in the model. Since very many models are tested by this strategy, it may be prone to overfitting and the "best" combination of variables must be validated very carefully (see Chapter 13).

Cross-validation or prediction testing can be used here the same way as was described in Section 7.2.3.

### 7.3.2 Forward selection, backward elimination and stepwise procedures

Instead of searching all subsets up to a given size, a more economical strategy is to find the best single variable, the best one to add to it, the best one to add to these two, and so on. This strategy is called *forward* selection. This is much easier computationally, but has the disadvantage that it is by no means guaranteed to find the best combinations. There is no reason why the best three variables have to include either of the best two. In NIR spectroscopy for example, all single variable equations will usually give very poor fits, and the choice of first variable can be almost random. Keeping this variable in all subsequent selections may very well cause us to miss the best pair, best three, and so on. Modifications of the procedure are made to allow the deletion of variables that do not seem to be contributing. This is often called stepwise multiple linear regression (SMLR). There are about as many variations on the latter as there are computer programs for stepwise regression.

Another strategy is to use *backward* elimination. Instead of starting with one variable and building up the equation to a reasonable size, this procedure operates the other way round. It starts with all variables and deletes uninteresting ones successively. This method can only be used in its standard form if the initial number of variables is smaller than the number of samples in the calibration set.

For all these methods, the F-test is probably the most frequently used for judging the importance of variables. For each step, the F-test compares the larger model with the smaller and provides a *p*-value measuring the degree of difference between the two models. For the forward procedures, new variables are incorporated as long as the new variable is significant. For backward procedures, variables are deleted until all remaining variables are significant.

As an illustration of the use of some of these methods, the NIR meat data set used in Chapter 3 for illustrating the selectivity problem is used. The data are based on 100 equidistant wavelengths in the NIR

**7.3** Search strategies

region between 850 and 1050 nm. A total of 70 samples are used for calibration and 33 are used for prediction testing. The focus here is on prediction of protein percentage. Using F-tests with a significance level of 0.05, the stepwise and forward selection methods gave the same results. They both ended up with five wavelengths in the solution, quite evenly spread over the whole wavelength region. The prediction ability measured by *RMSEP* (see Chapter 13) was equal to 0.52 with a correlation between measured and predicted *y*-value equal to 0.97. With significance level equal to 0.15, the two methods gave different results, stepwise resulted in seven variables and forward in nine variables. The correlations between measured and predicted *y*-values for both solutions were equal to 0.97. The *RMSEP* for the stepwise method was equal to 0.52 and for the forward selection method it was equal to 0.57. For all these solutions, MLR was used for estimating regression coefficients. Using PLS on the nine wavelengths selected by forward selection resulted in a *RMSEP* as low as 0.46 using seven PLS components. This indicates that the two strategies of variable selection and PLS can favourably be combined. If we compare these results with the PLS used for the full set of 100 variables, the corresponding *RMSEP* was equal to 0.54 (seven components), indicating that deleting variables before PLS can sometimes lead to improvements in prediction ability. It also indicates that variable selection combined with LS can sometimes be an alternative to regular PLS/PCR from a prediction point of view.

### 7.3.3 Stochastic search

Stochastic methods [Davis (1987)], i.e. methods involving some element of randomness, are used to tackle a variety of optimisation problems. With a well-defined criterion to optimise, and a space of possible solutions that is too large to search exhaustively, then one solution is to move around it randomly. Not completely randomly, but randomly enough not to get stuck in one corner just because one started there. Genetic algorithms is one such group of methods that has shown some promise for variable selection [see, for example, Davis (1987) or Leardi *et al.* (1992)].

### 7.3.4  Selecting variables using genetic algorithms

#### 7.3.4.1  Genetic algorithms

The idea behind genetic algorithms is to "copy" evolution, where random variations in the genetic makeup of a population combined with selection of the fittest individuals lead to progressive improvements to the breed. A closer analogy would be with the sort of intensive breeding programmes that have developed both the plants and animals used in present day agriculture. The selection pressures here have been deliberate and intensive, but the genetic changes exploited are still essentially random. The essential ingredients are some way of coding the candidate solutions to the problem so that they can produce offspring and mutate in some natural way, and some measure of the fitness of any candidate solution, so that we can select the fittest.

For the variable selection problem the natural way to code the solutions is to use binary strings as shown in Figure 7.1. Each possible subset of variables can be represented as a string of 0s and 1s, with a 1 in position $i$ if the $i$th variable is present in the subset and a 0 if it is not. The length of the string is the number of variables available, e.g. 19, 33, 100, 700. The number of 1s in any particular string is the number of variables in the subset represented. This will typically be quite small in successful subsets.

010010100001100000

**Figure 7.1. Illustration of coding in genetic algorithms.** A binary string representing a subset of 18 wavelengths that includes the second, fifth, seventh, twelfth and thirteenth wavelengths only.

Choose two parents
Parent 1: 010010100001100000
Parent 2: 000010011000001000

cut and crossover
0100101......000001100000
0000100......110000001000

to give two offspring
Offspring 1: 010010111000001000
Offspring 2: 000010000001100000

**Figure 7.2. Illustration of combination.** Two parents create two offspring with a single random crossover.

*7.3.4.2 Combination*

Now suppose we have a collection of such strings, making up a population of, say, 100 possible subsets, or solutions to the problem. To breed new solutions we select two individuals, i.e. two strings, and combine them to produce two new solutions. Various rules for combination have been proposed. A simple one is to cut both strings at the same randomly selected point and cross them over, as in Figure 7.2. It is possible to use more crossover points, right up to the limiting case where we decide to cross or not at each variable. Whatever combination rule is used, the offspring will be different from the parents, in a way that involves some randomness, yet they will resemble them in some respects. At least some of the offspring of good solutions ought also to be good solutions, and some of them could be expected to be better ones.

Before: **01001010001100000**
After:  **0100101001011100000**

**Figure 7.3.** Mutation in genetic algorithms. A random mutation—the tenth wavelength is added to the subset.

*7.3.4.3 Mutation*

If the population of solutions evolves only through combination as described, none of the offspring will ever include variables that are not present in at least one member of the initial population. To enable other variables to be tried we need to introduce some further randomness in the form of mutation. Having produced an offspring, we allow each of the bits in the string a small probability of changing randomly from 0 to 1 or 1 to 0. Typically the mutation probability might be chosen so that the average number of mutations per offspring is of the order of 1, and we will see the sort of mutations illustrated in Figure 7.3. The idea is to allow extra diversity without destroying the patterns inherited from the parents.

*7.3.4.4 Fitness*

To select good individuals we need a measure of fitness. In this case some assessment of the predictive performance of the equation is what is required. This might be obtained by fitting the equation to some calibration samples and then calculating a sum of squared prediction errors on a test set (see Chapter 13). The fittest individual is the one with smallest sum of squared prediction errors. Using a sum of squared cross-validation errors would be an alternative. Any sensible fitness measure can be used, so long as we avoid ones such as residual standard deviation $\hat{\sigma}$ or $R^2$ (see Section 7.2.1) on the calibration set that will excessively reward large subsets.

**7.3 Search strategies**

One can also tailor the fitness measure to the requirements of the individual problem. For example, a simple way of putting an effective upper limit of ten variables on the size of the subsets to be considered would be to add a large penalty to the fitness of any individual with more than ten 1s in the string.

*7.3.4.5 Selection*

It is usual to give individuals with good fitness ratings a higher chance of producing offspring than ones with poor fitness. One way would be to start with a population of, say, 100 individuals, then choose 50 pairs at random, thus producing a further 100 offspring. Next we reduce the 200 individuals to 100, keeping the fittest ones only, and iterate. Or we could reduce from 102 to 100 after each pair of offspring is produced. Another way of introducing selection pressure is to choose the parents with probability proportional to fitness, instead of using equal probabilities. Whatever approach is chosen, we allow the population to evolve as long as its fitness goes on improving, or until we run out of computer time. When the process is stopped, the final population should contain at least some subsets that perform well.

*7.3.4.6 Discussion*

The reader will hardly have failed to notice the scope for variations in the algorithm at all points: combination rules, mutation rates, population size, selection strategy, stopping rule, to name only the more obvious. These choices need to be tuned to the particular problem, and this seems to be more of an art than a science. At the time of writing there is enough published research on applying genetic algorithms to NIR variable selection to suggest that the approach has considerable potential for identifying good subsets in this application at least. However, it is certainly not a routine method as yet, and more work is needed before it becomes clear how best to tune these methods to the problem [see Leardi *et al.* (1992) for more detail]. As for the other methods above, it is important that results from genetic algorithms be validated properly.

There is no guarantee that a search of this type will find the best performing subset. As the natural world shows us, evolution can get trapped in local optima, the mammals of Australia being a good example. However, we do not necessarily need the best solution, we just need a good one. The kangaroo may not represent a globally optimal solution, but it works pretty well.

## 7.4 Using the jack-knife to select variables in PLS regression

The above discussion was mainly related to LS fitting of data to linear subset models. The idea was to select a small number of variables to be used in regular LS regression in order to solve the collinearity problem. Recently, there have been a number of attempts at combining variable selection with the use of PLS regression for model fitting. The idea is that PLS is often a more stable and reliable fitting method than LS, but some of the variables may still be uninteresting and unimportant both from an interpretation and prediction point of view. Publications describing ways of combining the two approaches in a successful way are Brown (1992), Martens and Martens (2000) and Westad and Martens (2000).

The method proposed in Westad and Martens (2000), so-called jack-knifing [see, for example, Efron and Gong (1983)], is of general importance. It is a technique for estimating variability of estimated parameter values, which can be used for many different types of models. In this context, the principle of jack-knifing is used for estimating the standard errors of the regression coefficient estimates in the PLS model. Such estimates of precision for PLS regression coefficients are very difficult to compute analytically, but the empirical technique of jack-knifing makes this possible in an easy way. The regression coefficients themselves can then be divided by their estimated standard errors to give t-test [or equivalently F-test, Weisberg (1985)] values to be used for testing the significance of the variables used in the model. The same procedure can be used for PCR or any other regression method.

The jack-knife is a very versatile technique, based on a similar principle to cross-validation (see Chapter 13.4). It deletes one sample (or several) at a time and the regression coefficients are computed for each subset. The set of regression coefficient vectors gives information about the variability and can be combined in a simple formula to give estimates of the standard errors.

In Westad and Martens (2000), the jack-knife method is used for backward elimination, but it can also be used for certain types of forward selection. It should, however, be mentioned that Westad and Martens (2000) discussed another and faster search strategy than the standard backward elimination method described above. Instead of deleting only one, they eliminated a number of variables at each step in the elimination process. All of them were deleted according to their jack-knife t-value. This procedure will speed up the search process considerably and may be a reasonable thing to do in, for instance, spectroscopy. The practical implications of this simpler strategy should, however, be investigated before a broader conclusion can be drawn. The results obtained in Westad and Martens (2000) were promising.

## 7.5 Some general comments

Using different strategies and criteria for the same data set can sometimes give confusing results. For instance, forward selection may suggest a totally different set of variables than backward elimination. This is just an aspect of the fact that there is seldom one model which is to be preferred over all the others. In many cases, there may be a number of choices which give approximately the same fit and prediction ability. In such cases, the interpretation aspects and prior knowledge should be used to select the most reasonable candidate.

The automatic procedures available in most commercial software are very valuable, but the user should not hesitate to override them when knowing better. If possible do not just look at the best, look at the good alternatives. There may be reasons for preferring one of them. For example, if an equation with one fewer variable is only just worse than the best it might be preferable because experience suggests that the fewer the variables the more stable the calibration is in the long run. If an equation that is good but not best makes obvious sense chemically whilst the best does not, then trust your judgement over that of the computer.

As was also indicated above, the variable selection methods search for optimal combinations of variables. Sometimes, very many alternatives are compared. Therefore, their apparent merits as measured on the calibration set may sometimes (especially with small data sets) be over optimistic. Therefore, it is a safe and good practice in this area to set aside an independent dataset to validate the predictor properly. In some cases, a rough way of eliminating the obviously uninteresting variables first, followed by a more detailed variable selection based on the interesting candidates may be preferred from an overfitting point of view. Using such strategies reduces the number of combinations to search through and may therefore be less prone to overfitting.

A number of comparisons of prediction performance of variable selection and full spectrum methods have been conducted [see, for instance, Almøy and Haugland (1994) and Frank and Friedman (1993)]. The variable selection methods can sometimes give very good results, but both from an interpretation and a prediction point of view it seems that methods like PCR and PLS are often to be preferred (a combination of the two strategies).

# 8 Data compression by Fourier analysis and wavelets

## 8.1 Compressing multivariate data using basis functions

As above, we assume that we have data that comprise a sequence of measurements $\mathbf{x} = (x_1, x_2, ..., x_K)^t$ taken at $K$ equally spaced points on a scale such as time, wavelength or frequency. Often there will be such a sequence for each of several cases or samples. One quite general idea that underlies several data compression methods is that of representing or approximating such a sequence as a weighted sum of basis functions. In general, each basis function $\mathbf{p}_a$ generates a sequence $p_{1a}, p_{2a}, ..., p_{Ka}$ and $\mathbf{x}$ is approximated by a weighted sum of these sequences, so that

$$\hat{x}_k \approx \sum_{a=1}^{A} t_a p_{ka} \quad (8.1)$$

for each $k = 1, ..., K$. The important feature of this sum is that the weight $t_a$ has only one subscript. It depends on which basis function generated the sequence, but not on which term in the sequence is being weighted. Thus we may express the approximation of $\mathbf{x}$ symbolically as

$$\hat{\mathbf{x}} \approx \sum_{a=1}^{A} t_a \mathbf{p}_a \quad (8.2)$$

This is essentially equivalent to (6.1) and the first equation in (5.1). A helpful way to visualise this is as curves being added together. If the sequence of $x$s is plotted against time, wavelength or whatever the ordering corresponds to, it will often look like a con-

---

**AA**
DWT: Discrete wavelet transform
FFT: fast Fourier transform
FT: Fourier transform
LS: least squares
NIR: near infrared
PCR: principal component regression
PET: polyethylene terephthalate
PLS: partial least squares

Note that the same symbols $\mathbf{p}$ and $t$ as in the description of the PLS/PCR model are used here. The reason is that they play the same role in the sense that the $\mathbf{p}$s act as basis vectors and the $t$s as coefficients to multiply the basis vectors by in order to obtain $\mathbf{x}$. See also below.

tinuous curve. Similarly, the sequence corresponding to any basis function, when plotted in the same way, will be a curve. The basis curves are added together, in appropriate proportions, to approximate **x**. Appropriate proportions here usually means that the coefficients $t_a$ are chosen to make the approximation to **x** as close as possible in a least-squares sense. The point of the whole exercise is that for a fixed basis, the information in **x** can be summarised by the set of coefficients $t_1, t_2, ..., t_A$ required to construct its approximation. If it is possible to get good approximations with $A$ much smaller than $K$ we can exploit this for data storage, data visualisation, multivariate calibration and so on.

There are two main ways to choose a basis. Either we use a family of mathematical functions, sine and cosine waves or wavelets perhaps, or we can use a basis that is derived from our data, which is what happens in principal component analysis and partial least squares regression (Chapter 5). The advantage of using such data-dependent bases is that they are very efficient, giving good approximations for small values of $A$. The disadvantages are that they are relatively expensive to compute, that the basis can depend strongly on the particular samples chosen to generate it, and that noise in the data propagates into the basis itself, as well as into the coefficients. Using mathematical functions may give simpler and more stable representations, at the cost of needing more terms in the sum.

In most cases it simplifies the calculations, improves numerical stability and makes the loadings easier to interpret if the basis functions are orthogonal to each other (see Appendix A). This means that for any distinct $a$ and $b$

$$\sum_{k=1}^{K} p_{ka} p_{kb} = 0 \qquad (8.3)$$

8.1 Compressing multivariate data using basis functions

That is, if we multiply any two basis sequences point-wise and sum the result we get zero. Essentially what this condition does is to ensure that the separate basis sequences are really quite different from each other. This makes, for example, determining the proportions in which they have to be mixed to recreate **x** a simpler and numerically more stable task. The sine and cosine waves used in Fourier analysis (see below) have this property, as do principal components [see Chapter 5 and Martens and Næs (1989)]. The basis sequences are often normalised so that

$$\sum_{k=1}^{K} p_{ka}^2 = 1 \qquad (8.4)$$

i.e. the sum of squares of each basis sequence is 1. This means that the length of each basis vector is equal to 1, and it is a unit vector. A basis comprising orthogonal unit vectors is described as orthonormal.

Fourier analysis (see Section 8.2), uses sine and cosine waves at different frequencies as basis functions. These are global functions, in that they extend over the whole length of the curve being represented. If we need a high-frequency wave to represent a sharp feature, such as a sharp peak or a step, at one point in the curve then we are going to need several more even higher frequency waves to cancel it out elsewhere. Fourier series are good at representing curves that show the same degree of smoothness, or spikiness, throughout. They are not very efficient at representing curves that are smooth in places and spiky in others.

A more natural way to try to represent spectra that are actually made up of many overlapping Gaussian (or other type of) peaks would be to use an appropriate set of such peaks as a basis. Then the spectrum would be approximated by a weighted sum of Gaussian peaks of varying location and width. This has the advantage that the basis functions are local. If we need a sharp peak at some point, this need not in-

**8.1 Compressing multivariate data using basis functions**

troduce high frequency components into the rest of the fit. Unfortunately it has the major disadvantage that such a basis is not orthogonal. Peaks that overlap will have strong correlations, and the fitting procedure may become very unstable. Thus although this approach is feasible for curves that can be represented as the sum of a very small number of such peaks, it does not work so well for more complex curves.

The key idea of wavelets (see Section 8.3) is that a wavelet family provides an orthonormal basis that is local in nature. Each basis function is a little wave that is either zero or very small outside a limited range, so one can have the advantages of a local description of the curve, yet the family has orthogonality properties that make it easy to determine the best fitting combination.

## 8.2  The Fourier transformation

### 8.2.1  Summing sine waves

Figure 8.1 shows two sine waves, one of one cycle and the other of three cycles, with amplitude of one third. It is not difficult to imagine that the addition of these two waves will produce the slightly more complex wave/function in Figure 8.2. An even more complex example is given in Figure 8.3, which shows a series of sine waves with odd numbers of cycles from 1 to 15 with decreasing amplitudes. The sum of these functions is shown in Figure 8.4. If we extend the number of sine waves (Figure 8.5), then the sum would approach the shape of a square wave as shown in Figure 8.6.

If we can go from a sine wave to a square wave just by adding sine waves, a natural question is whether there is any limit to what shape we can fit? In 1807 Joseph Fourier provided an answer to this. He was able to prove that any continuous curve could be

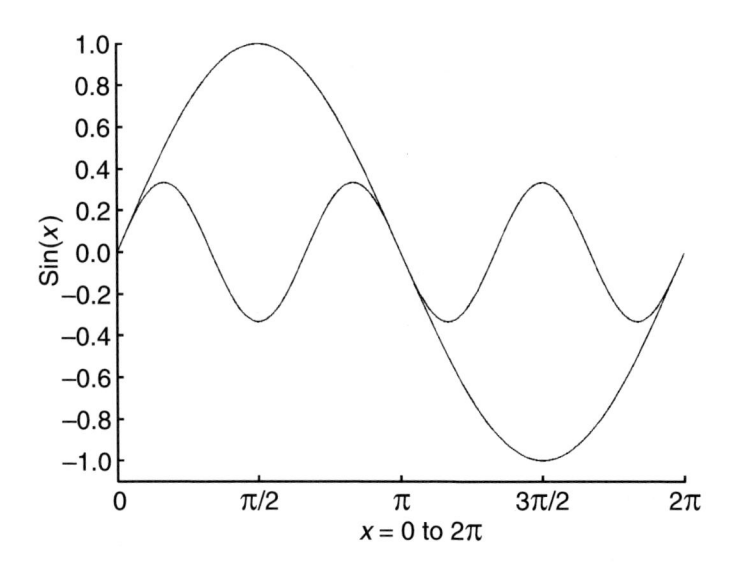

# Data compression by Fourier analysis and wavelets

**Figure 8.1.** Two sine waves.

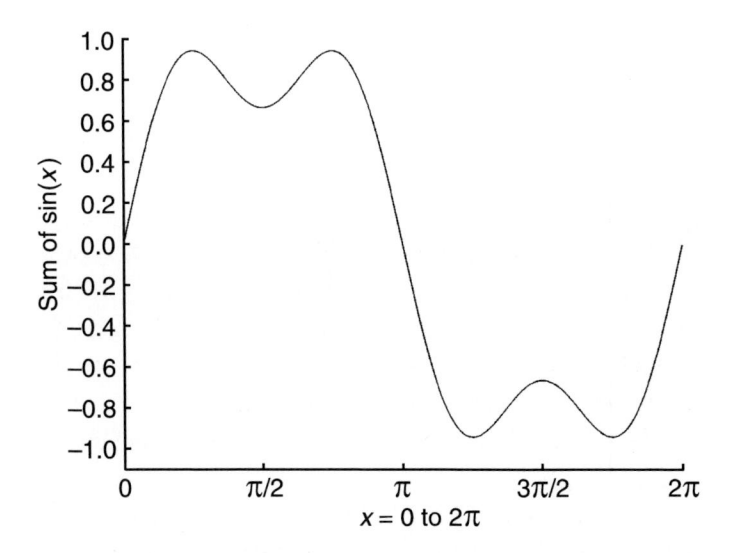

**Figure 8.2.** The sum of the two sine waves in Figure 8.1.

## 8.2 The Fourier transformation

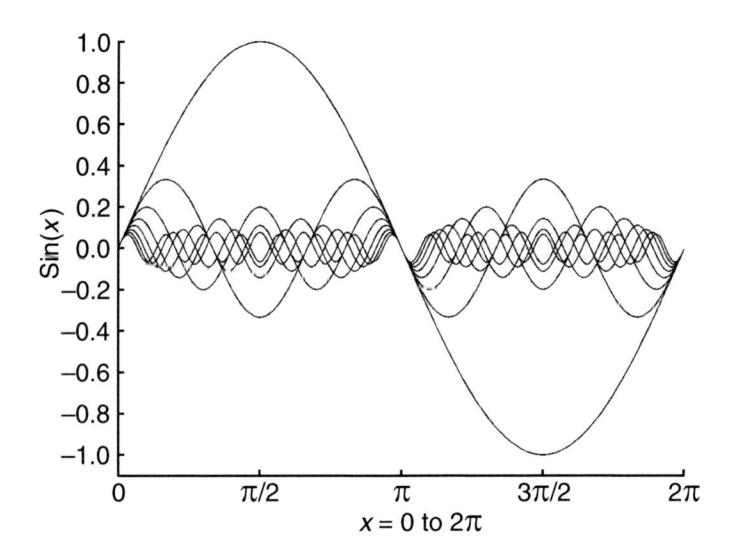

Figure 8.3. Eight sine waves of increasing frequency.

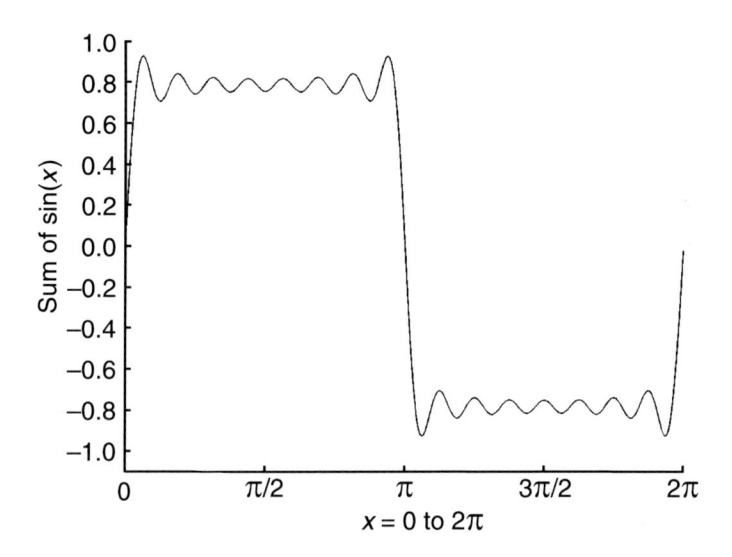

Figure 8.4. Sum of the sine waves in Figure 8.3.

8.2 The Fourier transformation

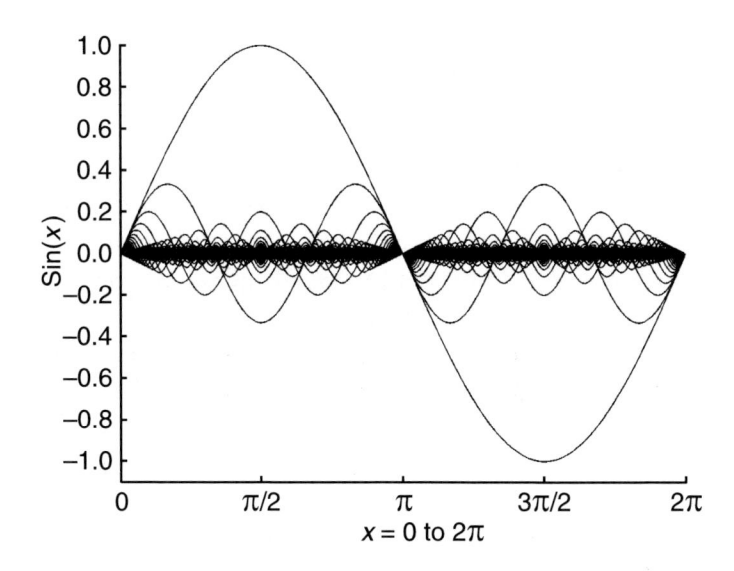

# Data compression by Fourier analysis and wavelets

**Figure 8.5.** One hundred sine waves of increasing frequency.

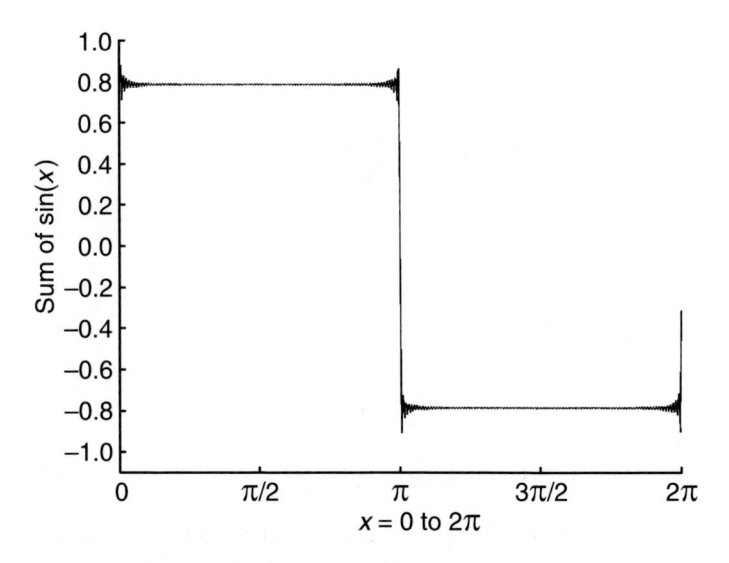

**Figure 8.6.** Sum of the sine waves in Figure 8.5.

## 8.2 The Fourier transformation

fitted by the summation of a series of sine and cosine waves and he provided the mathematics to do it. The determination of the sine and cosine components of a given function is known as the Fourier transformation (FT). The FT requires many computations and it was not until the fast Fourier transform (FFT) was discovered by Cooley and Tukey (1965) that even computers could manage to compute the FT of spectra with a large number of data points.

The idea of using the FT for the data processing of NIR spectra was described by Fred McClure and his colleagues [see, for example, McClure et al. (1984)]. If we have a spectrum that has been recorded at $K$ (usually equally spaced) wavelength intervals, $x_1$, $x_2$, $x_3$, ...., $x_K$, the FT [see, for example, Anderson (1971)] provides the decomposition:

> The formula (8.5) holds for $K$ even. If $K$ is an odd number, a slightly different formula must be used [see, for example, Anderson (1971)].

$$x_k = \sqrt{\frac{2}{K}} \left[ \frac{1}{\sqrt{2}} a_{01} + \frac{(-1)^k}{\sqrt{2}} a_{02} + \sum_{j=1}^{K/2-1} a_j \cos\left(\frac{2\pi j k}{K}\right) + \sum_{j=1}^{K/2-1} b_j \sin\left(\frac{2\pi j k}{K}\right) \right] \quad (8.5)$$

where the coefficients are given by

$$a_{01} = \sqrt{1/K} \sum_{k=1}^{K} x_k \quad (8.6)$$

$$a_{02} = \sqrt{1/K} \sum_{k=1}^{K} x_k (-1)^k \quad (8.7)$$

$$a_j = \sqrt{2/K} \sum_{k=1}^{K} x_k \cos\left(\frac{2\pi j k}{K}\right)$$
for $j = 1, 2, 3, ..., K/2 - 1$ \quad (8.8)

$$b_j = \sqrt{2/K} \sum_{k=1}^{K} x_k \sin\left(\frac{2\pi j k}{K}\right)$$
for $j = 1, 2, 3, ..., K/2 - 1$ \quad (8.9)

In other words, the spectrum $x = (x_1, x_2, x_3, ...., x_K)$ is written as a sum of sine and cosine basis functions with coefficients $a$ and $b$ determined by well-defined linear combinations of the *x*-data. The *a*s and *b*s are called the Fourier coefficients and determine how much the different frequencies contribute to the full signal. They therefore correspond to the *t*s in Equation (8.2). Note that the transformed signal represented by the *a*s and *b*s contain the same number of data points as the original spectrum. The (*a*, *b*) representation of the data is often called the "frequency domain" as opposed to the "time domain", the latter being defined as the original scale *x*. Note that the two first terms in the sum of (8.5) are equal to the average value of all the *x*s and half the average of all differences between adjacent values in **x**, respectively. For continuous spectra with *K* reasonably large, the latter will be small (in most applications, many of the terms will also cancel each other).

NIR spectra and many other chemical data often show an upward (or downward) linear trend. One of the assumptions behind the FT is that the waveform repeats to infinity, from each end. If the ends of the spectrum are different, the discontinuity will cause so-called "ringing". This means that rapid oscillations are generated at each end of the spectrum. Usually, one makes the two ends of the spectrum equal by joining them with a straight line and then subtracting it across the spectrum. Figure 8.7 is a NIR spectrum of a piece of a PET polymer, which was recorded at 2 nm intervals over the range 1100–2498 nm. In the figure it is shown before and after this linear correction, which is often referred to as "tilting". When reconstructing the spectrum from the Fourier coefficients, one has to remember to "un-tilt" it.

As can be seen from equation (8.5), an FT spectrum is composed of two series of *a* and *b* coefficients and there is no agreed manner of presenting this data in a graph. One way of presenting the coefficients is to show all the "*a*" coefficients followed by all the "*b*"s. Thus the FT of the NIR spectrum in Figure 8.7 is

**8.2 The Fourier transformation**

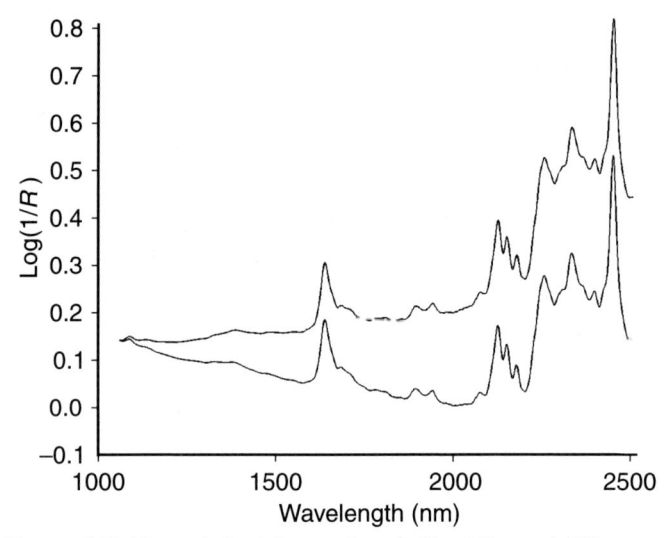

Figure 8.7. The original (upper) and tilted (lower) NIR spectra of a sample of PET.

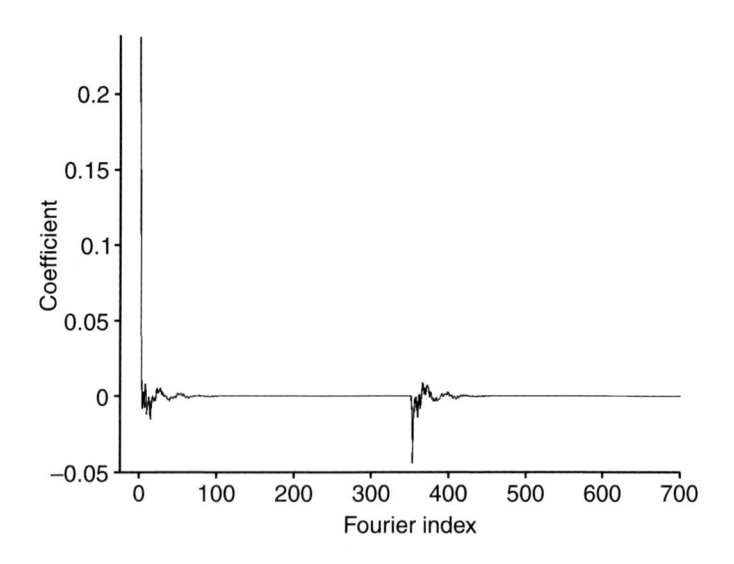

Figure 8.8. The Fourier coefficients for the spectrum of PET.

8.2 The Fourier transformation

shown in Figure 8.8, while a close up of some of the larger *a* coefficients is shown in Figure 8.9.

### 8.2.2 Data compression

It is often found that most of the useful information in the frequency domain is found at low to moderate frequencies and that the high frequencies can be ignored without losing too much information. In other words, many of the pairs of Fourier coefficients will be close to zero, as seen in, for instance, Figure 8.8. If we put the sine and cosine basis functions corresponding to the low frequency coefficients into a matrix $\hat{\mathbf{P}}$ and denote the corresponding Fourier coefficients by $\hat{\mathbf{t}}$, it is easily seen that one can write the approximation for a spectral matrix

$$\hat{\mathbf{X}} \approx \hat{\mathbf{T}}\hat{\mathbf{P}}' \qquad (8.10)$$

corresponding exactly to the data compression in equation (8.2). It follows directly from Chapter 5 how the compressed information collected in $\hat{\mathbf{T}}$ may be used in a regression context (see equation 5.4).

Fourier analysis is also frequently used for smoothing of spectral data. If high frequency contributions are ignored when reconstructing the spectrum using (8.5), this removes high frequency noise. Low frequency noise will still be present in the reconstructed data.

We refer to Figures 8.9–8.12 for an illustration of data compression and smoothing. In Figure 8.10 only 20 pairs of Fourier coefficients from the full FT above are used (see also Figure 8.9). The reconstructed spectrum gives a poor fit to the original as shown clearly by the difference spectrum at the bottom. If we use an adequate number of Fourier coefficients, the differences will be very small and are often seen to contain just noise. Figure 8.11 shows the result of using 100 pairs. The reconstructed spectrum ap-

> Hats, ^, are used since we have considered computed quantities (see also Chapter 5).

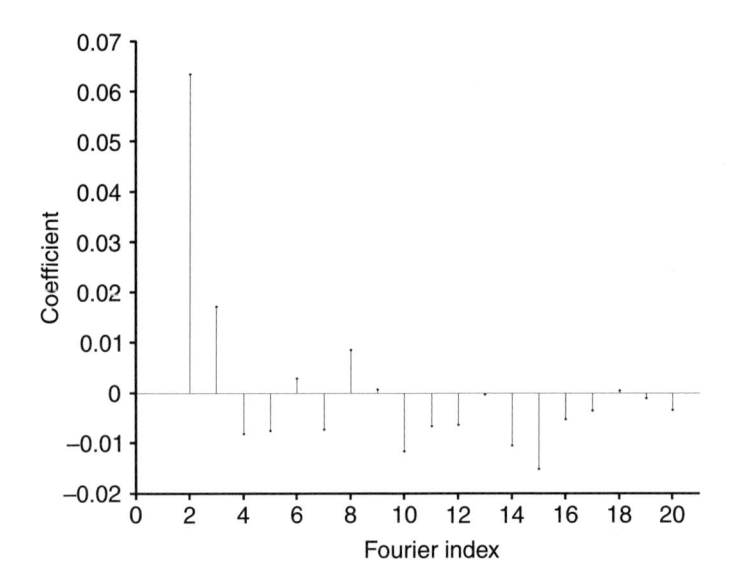

Figure 8.9. Some low frequency Fourier cosine coefficients from Figure 8.8. The first coefficient has been omitted to allow the others to be seen.

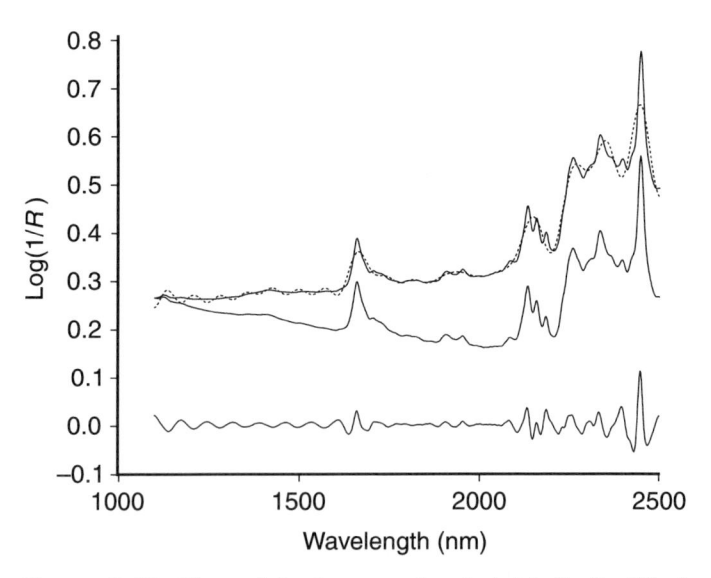

Figure 8.10. The original, reconstructed (dotted), tilted (middle) and difference (bottom) spectra using 20 pairs of Fourier coefficients for the reconstruction of the spectrum in Figure 8.7.

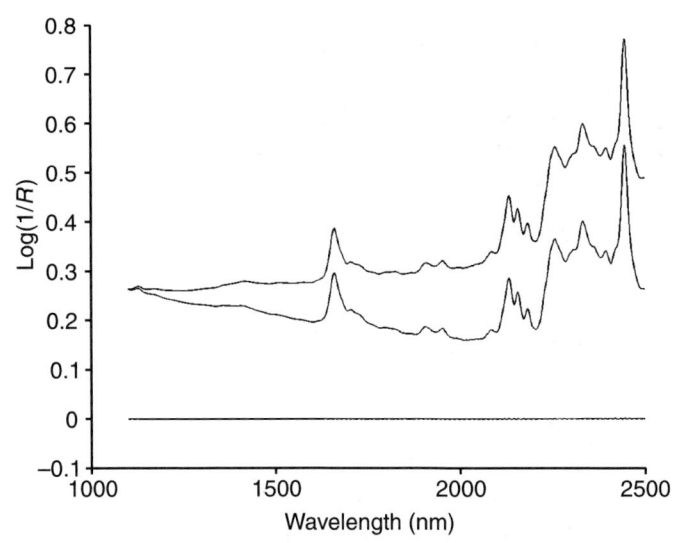

Figure 8.11. As Figure 8.10 but using 100 pairs of Fourier coefficients. The almost straight, horizontal line is the difference between the original and reconstructed spectra, which is expanded in Figure 8.12.

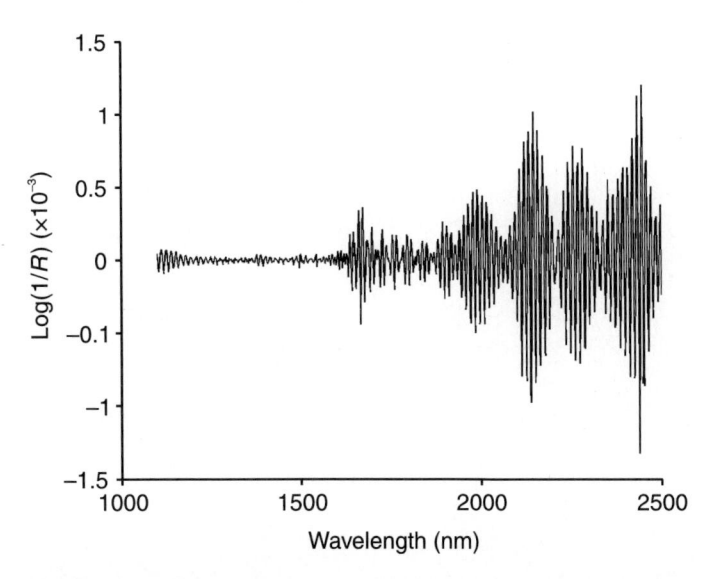

Figure 8.12. The difference spectrum plotted on its own scale using 100 pairs of Fourier coefficients for the reconstruction of the spectrum.

## 8.2 The Fourier transformation

pears to match the original perfectly and we have to look at the difference spectrum to see any errors. The errors are very small (Figure 8.12), but we have achieved a 70% reduction in the size of the data file. Thus the FT process can achieve data compression and noise reduction in one single operation.

## 8.3 The wavelet transform

### 8.3.1 Wavelets

Figure 8.13 shows some of the many possible examples of wavelets, each one of which can be used to generate a family that can be used as a basis for representing curves. The Haar wavelet, the square one in the first panel, has been around for a long time, but the rest are relatively recent inventions. The other three

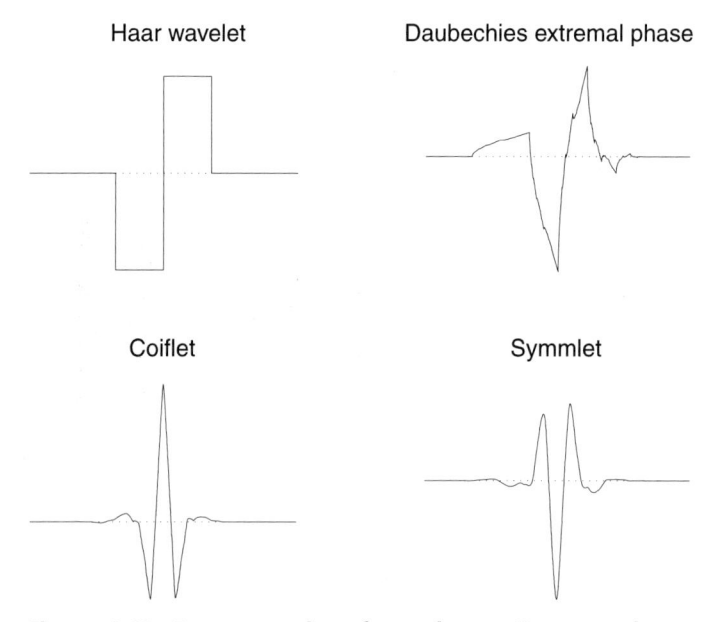

Figure 8.13. Four examples of wavelets, a Haar wavelet, a Daubechies extremal phase wavelet, a coiflet and a symmlet.

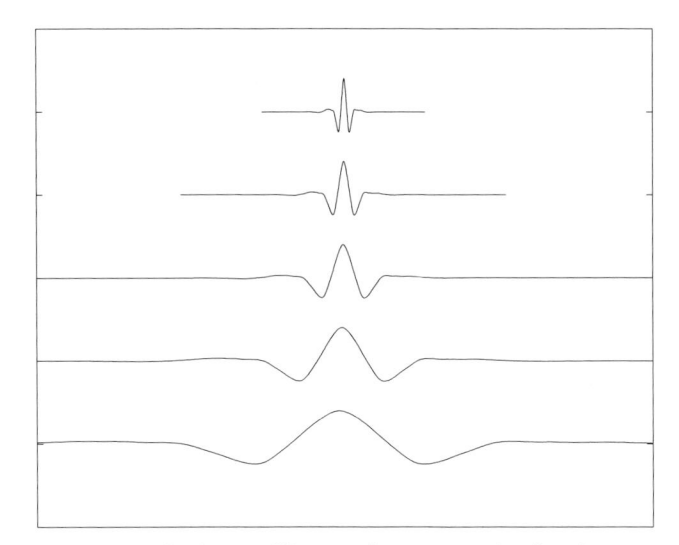

# Data compression by Fourier analysis and wavelets

**Figure 8.14.** Dilation: coiflets at five successive levels.

shown are due to Ingrid Daubechies (1992) and only date from the late 1980s. At first sight all these wavelets have some rather undesirable features. Only the Haar wavelet has symmetry, all the other three being asymmetric to some degree. The Haar and Daubechies "extremal phase" wavelets are also rather rough in shape. The coiflet and symmlet are attempts to find smooth, nearly symmetric wavelets that are exactly zero outside a limited range and generate an orthonormal family. This is a surprisingly difficult task. In fact extremal phase wavelets like the one shown are widely and successfully used, despite their somewhat unnatural appearance.

To create a family that can be used as a basis, we take one of these so-called mother wavelets and produce others by two operations, dilation and translation. Figure 8.14 shows dilation. The wavelet, a coiflet, on the lowest line is halved in width on the line above, halved again on the line above that, and so on. The top two lines show the full non-zero extent of the wavelet. The three largest wavelets have had their

**8.3** The wavelet transform

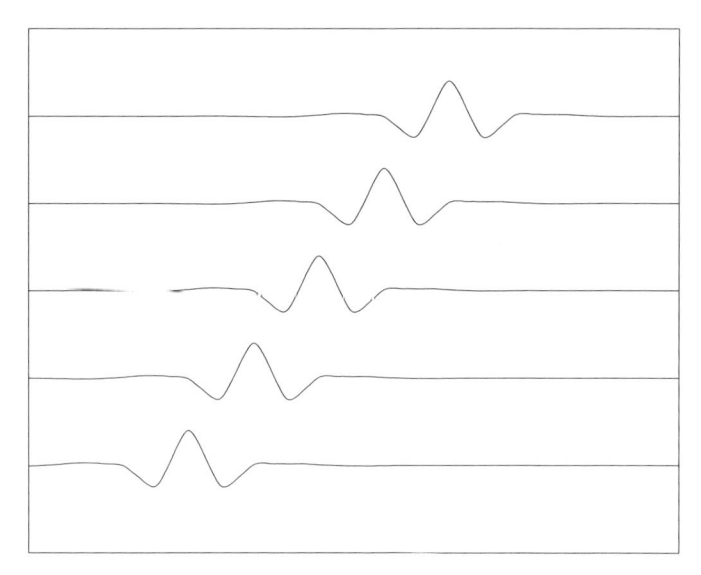

**Figure 8.15.** Translation: a sequence of five coiflets at the same level.

tails, where they are almost but not quite zero, cut off for display purposes. Figure 8.15 shows the other operation, translation. The same coiflet is repeatedly shifted to the right as we move up the sequence. The shift is carefully specified to make each of these wavelets orthogonal to each of the others. The full family comprises wavelets of a range of sizes, as in Figure 8.14, with a whole sequence of translated wavelets at each size level, as in Figure 8.15. The orthogonality property holds across levels as well as within levels, so that the wavelet family, suitably scaled, forms an orthonormal basis which can be used to approximate curves.

There are a few practical details to tidy up before we can use the wavelets to approximate curves such as measured spectra. The pictures show continuous wavelets that, in principle, continue off the top and bottom of Figures 8.14 and 8.15. That is, they can be as large or small as we like and at any level there is an infinite sequence extending in both directions. In practice any real curve is measured at a discrete number of points over a finite range. Thus we use discrete wavelets, evaluated at the same points as the curve being approximated, and we only use a limited range of levels, from one big wavelet that just fits the full range down to the smallest ones that make sense given the resolution of the curve. We also have to cope with end effects. Usually the curve is extended in some arbitrary way, either by padding it with zeros or extending it as a periodic function, so that some of the wavelets can be allowed to overlap the ends. All these details will be familiar to anyone who has met Fourier analysis. What would also be familiar is the existence of a fast algorithm, the discrete wavelet transform or DWT, that parallels the fast Fourier transform (FFT) and works best when the number of data points is a power of two.

Having highlighted the parallels with Fourier analysis it is perhaps worth remarking again on the difference. There is an obvious correspondence between the different levels of the wavelet decomposition and the different frequencies of a Fourier decomposition. At any particular level (or frequency), however, the wavelet family has a whole sequence of local basis functions, each of which may or may not contribute to the approximation of the curve. This is in contrast to the Fourier basis functions which are global, and will contribute over the whole range of the curve if given a non-zero weight.

### 8.3.2 An example

Figure 8.16 shows an artificial "spectrum" created as a mathematical function by Donoho and Johnstone (1994) and used by these and other authors as a test example. It is discretised at 512 points, so that in the notation introduced in Section 8.1, $K = 512$. Figures 8.17 and 8.18 illustrate a wavelet decomposition

Figure 8.16. An artificial spectrum.

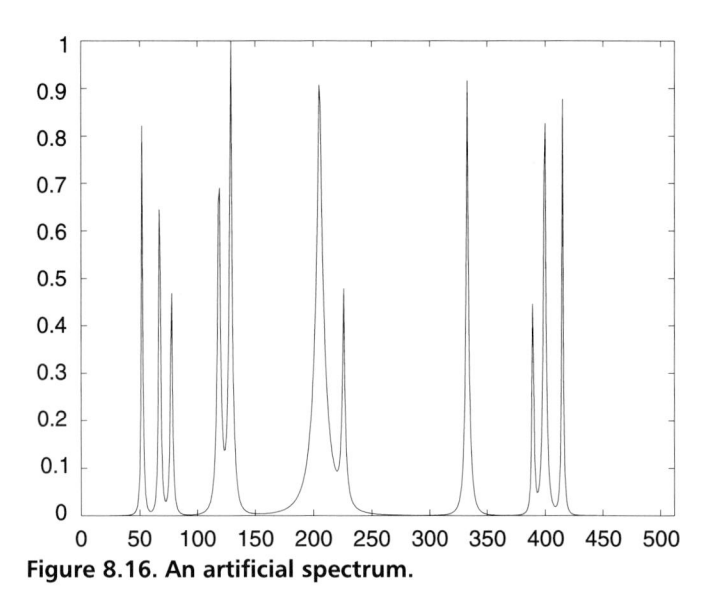

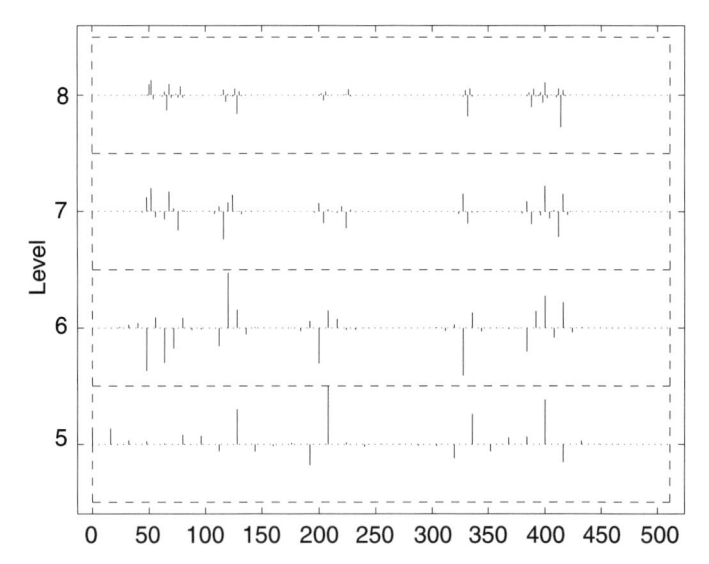

Figure 8.17. Coefficients at the four finest levels of a wavelet decomposition of the curve in Figure 8.16.

of this curve, using symmlets. In Figure 8.17 are plotted the coefficients, the $t_a$s in Equation 8.1, of the wavelets at the four finest levels, i.e. starting with the

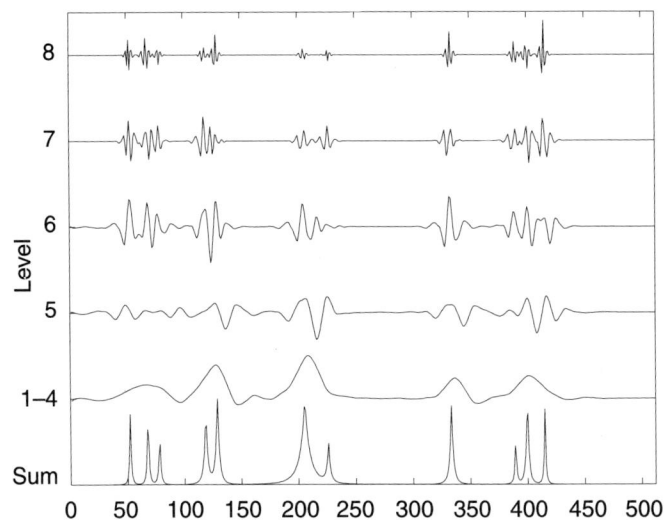

# Data compression by Fourier analysis and wavelets

**Figure 8.18.** Wavelet decomposition of the curve in Figure 8.16.

smallest wavelets on level 8. There are 256 of these smallest wavelets, 128 at level 7, 64 at level 6, 32 at level 5 and so on. Note that the only differences between wavelets at the same level are that they are positioned differently. The dotted line on each level of the plot is zero, so bars going upwards represent positive coefficients, bars going downwards negative ones. If we take the levels one at a time, and for each level multiply the coefficients by the wavelets and sum, we get the representation in Figure 8.18, called a multi-resolution analysis. This shows the contributions made by the wavelets at each level (levels 1–4 have been combined) to the reconstruction of the full signal shown at the bottom of the plot. What we can see from this analysis is not only the contributions at different frequency levels, something one could get from a Fourier analysis, but also where along the length of the signal the different frequencies contribute.

This reconstruction is exact. In the full decomposition there are 512 coefficients of 512 basis func-

8.3 The wavelet transform

tions, and these coefficients carry all of the information in the 512 data points in the original curve. The 512th basis function (256 + 128 + ... + 2 + 1 = 511) is a differently shaped curve called a scalet or father wavelet that is a matching partner (in some sense) to the mother wavelet. Just one of these, at the coarsest level, is needed to complete the set.

It might appear that we have achieved very little, replacing 512 data points by 512 coefficients. However, if we look at the coefficients in Figure 8.17, we see that only about 50 of the 256 at level 8 are visibly non-zero, as are about 30 of the 128 at level 7. Thus we could probably throw away more than half of the coefficients and still get a very good reconstruction of the signal. With smoother, less spiky signals, we might be able to discard the whole set of level 8 coefficients and still get a good reconstruction, perhaps even a better reconstruction than the original because it has lost some high frequency noise. There is a lot of literature, Donoho and Johnstone (1994) being a good, if somewhat technical example, on what is called thresholding, which usually means setting small wavelet coefficients (meaning small coefficients, not necessarily the coefficients of small wavelets) to zero according to a variety of schemes, in order to remove noise from a signal.

### 8.3.3 Applications in chemistry

A comprehensive review paper [Leung *et al.* (1998)] lists more than 70 papers applying wavelets to problems in chemistry. The web site of the leading author of this review paper (A.K. Leung) listed 230 such papers by March 2000. Areas of application include chromatography and mass spectrometry, as well as most forms of spectroscopy. The main uses for wavelets are for smoothing, noise removal and data compression [Leung *et al.* (1998), Walczak and Massart (1997), Alsberg *et al.* (1997)]. The smoothing and noise removal are pre-processing steps, aimed at improving signal quality.

The data compression may be for efficient storage, or as a data reduction step after which the wavelet coefficients (scores), or at least some of them, are input to a regression or discriminant analysis procedure (see Chapter 5 and Chapter 18). Note that this type of regression is very similar to what is done for PCR and PLS. First, scores are computed, then these scores are used in a LS regression (see Chapter 5). The only difference is the way the scores are computed. For the wavelet method, the scores come directly as coefficients for mathematically defined functions, while for PCR and PLS they are obtained using loadings computed from the data.

Wavelets have been or are about to be used for everything that Fourier analysis would have been the natural tool for a few years ago. In addition to the specifically chemometric review [Leung et al. (1998)] and tutorial [Walczak and Massart (1997) and Alsberg et al. (1997)] papers already cited, there are many books on wavelets. Most of these involve some quite technical mathematics. The classic book by Daubechies (1992) is very technical, those by Chui (1992) and Chui et al. (1994) perhaps a little less so. The book by Ogden (1997) is more readable.

## 8.3 The wavelet transform

# 9 Non-linearity problems in calibration

In the chapters above we have primarily described linear calibration techniques, i.e. methods that combine the spectral variables in a linear prediction formula (see, for example, Equation 5.2). Such methods are definitely the simplest, both theoretically and in practice and surprisingly often they work well. In some cases non-linear methods can, however, give substantial improvements [see, for example, Næs et al. (1990)].

## 9.1 Different types of non-linearities exist

First of all it is important to emphasise that there are different types of non-linearities [see Martens and Næs (1989)]. Look for instance at Figure 9.1 (identical to Figure 6.2), which is an illustration of the relationship between two $x$-variables and one $y$-variable. In this example, there are strong non-linearities between each of the spectral variables $x_1$, $x_2$ on one side and $y$ on the other. These are called univariate non-linearities. On the other hand, when $y$ is looked upon as a function of both $x_1$ and $x_2$, the relationship is nicely linear. In other words, even though any single plot between a pair of variables is non-linear, the multivariate relation is linear.

Such non-linearities are therefore not so important in practice since they can be modelled by the linear calibration methods discussed above. More problematic are those non-linearities that cause a totally non-linear relationship between $y$ and all the $x$-

LS: least squares
LWR: locally weighted regression
MSC: multiplicative scatter correction
NIR: near infrared
OS: optimised scaling
OSC: orthogonal signal correction
PCA: principal component analysis
PCR: principal component regression
PLS: partial least squares

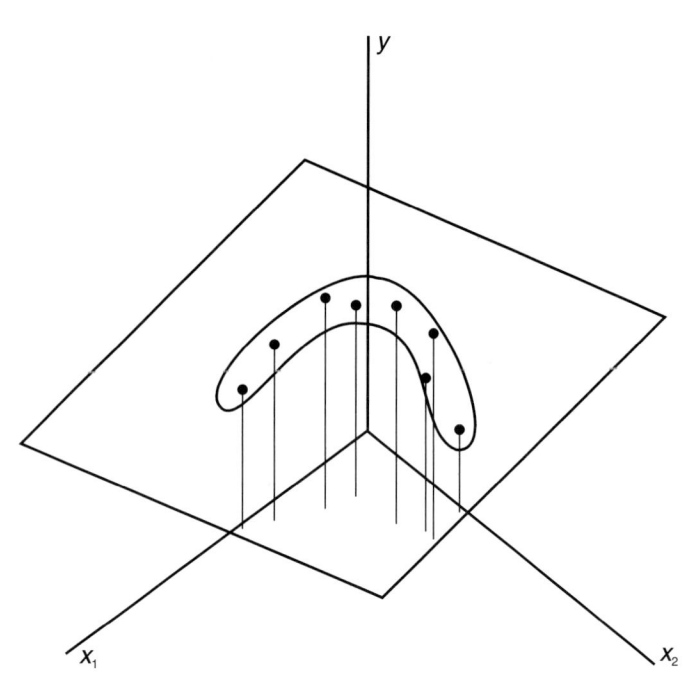

Figure 9.1. In this example there are strong non-linearities in the relationships between any pair of variables. The variable y can, however, be written as a linear function of $x_1$ and $x_2$ when they are modelled simultaneously.

variables simultaneously. These are called multivariate non-linearities. Such regression surfaces can never be modelled as linear functions of the $x$-variables and alternative methods must be found in order to obtain satisfactory prediction results.

It is, however, important to mention that situations exist where the individual non-linearities, even when there is multivariate linearity (Figure 9.1), are important enough to favour a non-linear approach. The reason for this is the following: in general, when a small number of calibration samples is available, it is important for the stability and reliability of the predictor that the number of free parameters to be estimated is limited. For PLS and PCR, this means that the number of components in the model should be as small as possible. For instance, in the situation considered in Figure 9.1, the $y$ variable is not only a linear function of both $x$-variables simultaneously, but also a non-linear function of the first principal component of the $x$-variables. In such cases a good non-linear model based on a small number of principal components can give a more stable and reliable calibration equation than a linear function of a larger number of principal components incorporated in order to account for non-linearities. Therefore, even non-linearities of the simple type may be better handled by non-linear methods than the regular PLS/PCR approach.

## 9.2 Detecting multivariate non-linear relations

The most important tool available for detecting non-linearities is the residual plot. This is a plot of the $y$-residuals $\hat{f} = y - \hat{y}$ versus $\hat{y}$ or versus other characteristics of the data, for instance sample number or principal component direction [see Weisberg (1985) and Martens and Næs (1989)]. Note that plotting of $y$-residuals versus $\hat{y}$ is strongly related to plotting $y$ versus $\hat{y}$, so clear non-linearities detected by one of these techniques will also usually be detected by the other. The regular residual plot is usually to be preferred.

There are, however, cases where outliers are difficult to detect when using the standard plots. Look for instance at the PCR residuals in Figure 9.2. This is an example based on NIR measurements of mixtures of fish protein, water and starch and the calibration is for protein percentage. In this case, there are strong non-linearities in the residuals, illustrated by the light and dark regions, corresponding to positive and negative residuals, respectively. From the $y$ versus $\hat{y}$ plot in Figure 9.3 it is, however, impossible so see that such a non-linearity is present. The reason for this is that the

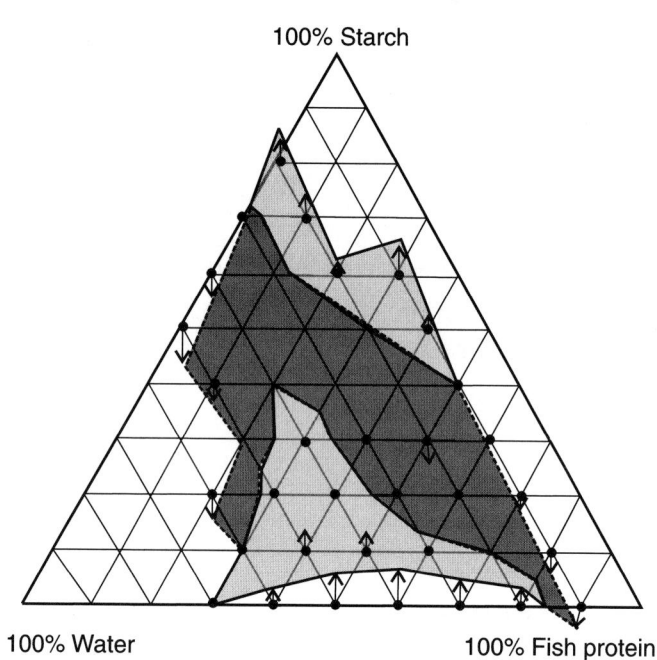

**Figure 9.2.** Prediction residuals presented in the experimental design region. Some arrows are incorporated in order to indicate the sign of the residual (+/–). The light and dark shaded regions show where the residuals are positive and negative, respectively. The same data were used in Figure 6.1. Reproduced with permission from T. Isaksson and T. Næs, *Appl. Spectrosc.* 42, 7 (1988) © Society for Applied Spectroscopy.

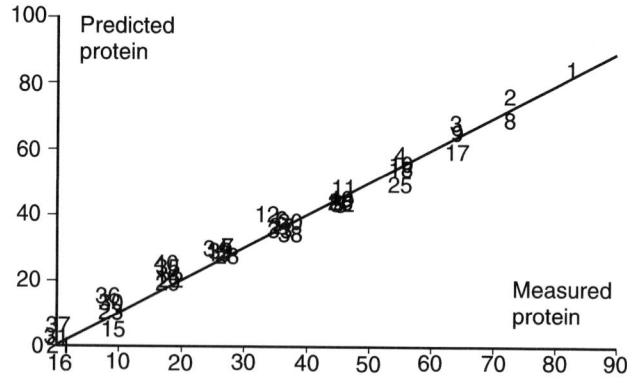

**Figure 9.3.** Plot of $y$ versus $\hat{y}$ for the PCR calibration whose residuals are plotted in Figure 9.2. The data are from protein predictions based on NIR spectra using mixtures of fish protein, water and starch.

non-linearity in Figure 9.2 is "orthogonal" to the protein direction ($y$) in the plot. The consequence is that the non-linearity is interpreted as "noise".

In such cases, plotting versus other directions in the data-space (for instance principal component directions) can be useful. In some cases, however, the only possibility is simply to try non-linear techniques and compare them with a linear approach using some sort of validation method (Chapter 13).

We refer to Miller (1993) for a discussion of the spectroscopic basis for non-linearities.

## 9.3 An overview of different strategies for handling non-linearity problems

Figure 9.4 gives an overview of different strategies for solving non-linearity problems in calibration.

An important class of methods is the transformation techniques. These are methods that transform the data prior to linear calibration. Such transformations can be applied to $y$, to $\mathbf{x}$, or to both, they can be based on statistical fitting procedures or founded on spectroscopic theory. Another common way of solving non-linearity problems is by adding extra variables, obtained as functions of the original $x$-variables. A third possibility is to use a non-linear regression technique.

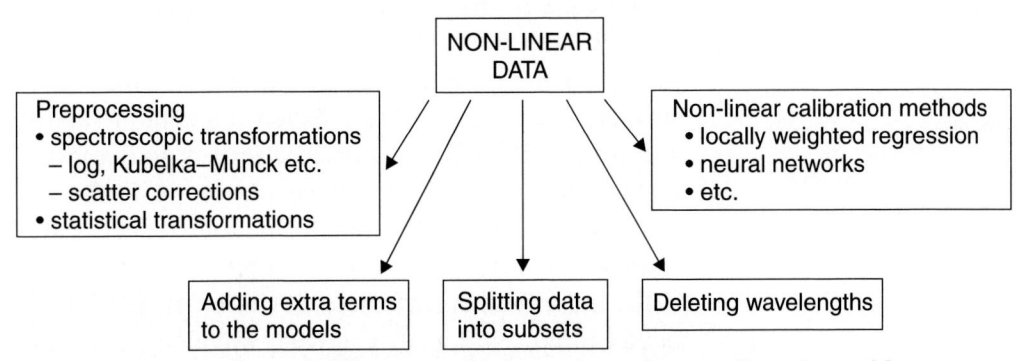

**Figure 9.4.** An overview of different methods used to solve non-linearity problems.

This is a rich family of methods, some of which are useful and easy to use and some of which are complicated and not so well suited for practical calibrations. For instance, the parametric non-linear regression methods, which assume a specified parametric relationship between the variables, are difficult to use in NIR spectroscopy, because of the complexity of the NIR data. Non-parametric techniques based on, for instance, local averaging are flexible and can sometimes be very useful. Likewise, so-called semi-parametric techniques, for instance neural networks and locally weighted regressions, may also be useful.

In the following sections we will discuss briefly each of these groups of methods. In Chapters 10, 11 and 12 we will go into more detail for some of the techniques.

### 9.3.1 Spectroscopic transformations

The most frequently used transformation in spectroscopy is the log function applied to the spectral absorbances. The basis for this transformation is derived from Beer's Law (1852), which states that the concentration of an absorbing species, in a non-scattering solution, is related to the logarithm of the relative transmitted intensity of the incident, monochromatic radiation. If we measure a percent transmittance, $T$, then the relative intensity will be $1/T$ and in the case of diffuse reflection measurements, $1/R$. Even though Beer's Law is difficult or impossible to justify in many applications, this transformation is still used more or less as a standard technique and it seems to work well in many situations. Another transformation much used in IR spectroscopy is the so-called Kubelka–Munck transformation [Kubelka and Munck (1931)]. This is also a simple method applied to the individual absorbances. It was originally developed to improve linearity at each wavelength for turbid samples. It is important to emphasise that both

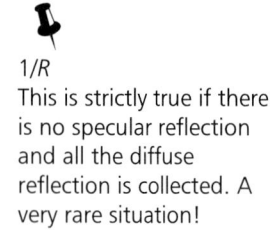

$1/R$
This is strictly true if there is no specular reflection and all the diffuse reflection is collected. A very rare situation!

these transformations are essentially univariate, and, therefore, they do not necessarily improve the multivariate linearity (see Section 9.1).

Another class of methods for spectroscopic pre-processing is the family of so-called scatter or signal correction techniques. These are methods that try to eliminate or reduce the effect of light scatter in diffuse spectroscopy. Multiplicative scatter correction [MSC, Geladi *et al.* (1985)], sometimes also called multiplicative signal correction, is one such technique which has shown to be very useful for improving multivariate linearity and prediction ability. For each sample the MSC method is based on regressing the spectral readings onto the corresponding values of the average spectrum. The estimated coefficients are then used in order to correct for differences in scatter among the samples. The method has later been modified in different ways to account for known interference effects [Martens and Stark (1991)] and different scatter-level at different wavelengths [Isaksson and Kowalski (1993)]. Another way of reducing scatter effects is by using derivatives (in fact derivatives remove additive rather than multiplicative effects). Optimised scaling (OS) developed by Karstang and Manne (1992) is another related technique which has shown promising results. It is based on optimising both the absorbance levels and the calibration constants simultaneously. Orthogonal signal correction (OSC) is a new technique put forward by Wold and co-workers (1998).

### 9.3.2 Statistical transformations

Several statistical transformations are available [see, for example, Weisberg (1985)]. Statistical transformations can be used for both $y$ and $\mathbf{x}$. If a transformation of $y$ (for instance, fat content of a sausage) is used, the predicted values have to be transformed

**9.3 An overview of different strategies for handling non-linearity problems**

back to the standard scale before they can be used in practice.

The most well-known statistical transformation to solve multivariate non-linearity problems is probably the Box–Cox transformation [see, for example, Weisberg (1985)]. This method is based on a parametric model (power model) for the transformation of $y$. The model contains a transformation parameter that can be estimated from the data. The estimation is done in such a way that linearity between $y$ and linear functions of **x** is optimised. The method has been much studied and has also been extended in different ways. As far as we know, no papers have been published with this transformation in NIR spectroscopy. It deserves to be given a try.

### 9.3.3 Adding extra terms to the model

Another possibility is to add extra terms to the "$x$-side" of the mathematical model. The most common terms are squares and products (interactions) of the $x$-variables themselves [Berglund and Wold (1997)]. A similar approach has also been tested on the principal components [see, for example, Næs and Isaksson (1992a)]. The collinearity is then solved by the PCA and the non-linearity is solved by adding squares and interactions. This method has been extended by Oman *et al.* (1993) where a graphical technique was presented for picking the most interesting terms to add.

### 9.3.4 Splitting data into subsets

In some cases it is possible to split non-linear data into subsets in such a way that linearity is acceptable within each of the subsets. Several possibilities exist for this, but methods based on cluster analysis (see Chapter 18) are among the most appealing. A modification of straightforward cluster analysis for this purpose was proposed by Næs and Isaksson

(1991). The cluster criterion was defined as a weighted average of a Mahalanobis distance in $x$-space and the $y$-residual $\hat{f}$. The method ensures that samples collected in one cluster are both close in $x$-space and exhibit a fairly linear relationship between **x** and $y$ within each cluster. Næs (1991) investigated how a prediction should best be performed in such cases.

### 9.3.5 Non-linear calibration methods

Non-linear methods can be parametric, non-parametric or something in between, so-called semi-parametric. Parametric methods are based on a specified model with a few unknown parameters, which are usually estimated using least squares (LS, see Appendix A). Such methods are, however, very difficult to apply in, for instance, NIR spectroscopy where the number of variables is usually very large, and where lack of theoretical knowledge makes it very difficult to set up such a parametric model. In practice, it is usually more useful to apply a non-parametric approach, i.e. an approach that is not based on a clearly defined parametric relation, but is based on statistical smoothing or averaging principles instead. The only assumption needed for such cases is that the surface relating $y$ to the $x$s is reasonably smooth. The available techniques try to balance between flexibility on one side and parsimony (few parameters) on the other.

Before discussing a few of the most promising approaches to non-parametric or semi-parametric model fitting, it is important to mention that it is impossible to find one single non-linear method which is best for all possible cases. It is always possible to find an application where each of the methods works better than most other methods. What is possible, however, is to find out which of the different approaches looks most promising within the actual application one is interested in, for instance, NIR spectroscopy.

**9.3** An overview of different strategies for handling non-linearity problems

### 9.3.6 Locally weighted regression (LWR)

Locally weighted regression is based on the idea of local linearity and is a quite straightforward extension of PCR [see, for example, Næs *et al.* (1990)]. First, the principal components are computed. Then for each new prediction sample, the spectrum **x** is projected down onto the first few principal components and the calibration samples which are closest in this low-dimensional space are identified. Then a standard least squares solution is obtained using the local samples and the few principal components only. Thus, a new calibration is performed for each new prediction sample based on a local subset of the calibration samples. If the number of samples in each local calibration and the number of principal components required to obtain a good result are both small, this represents no problem from a computational point of view. The method has been tested in several applications of NIR spectroscopy [and also in other cases, see Seculic *et al.* (1993)], and it seems to work well. In some cases it has represented an improvement of up to 30 to 40% compared to standard linear methodology. Another advantage of the method is that, since it is based on principal components and standard least squares fitting only, it is easy to visualise and understand, and all diagnostic tools available for linear methods are also available here. The method has been extended and modified in different directions in order to account for different types of non-linearities (see Chapter 11).

The LWR methods are quite closely linked to kernel methods for local averaging of samples. Such methods have been tried in, for instance, fluorescence spectroscopy [see, for example, Jensen *et al.* (1982)].

### 9.3.7 Neural networks

"Neural networks" is a family of techniques that can be used for many types of applications. In the context of calibration, however, we confine ourselves to the so-called feed-forward networks. These are networks that "feed" the information from the input layer, i.e. input data (**X**), into an intermediate or hidden layer, before the hidden variables are combined to give the output variable(s) (**y**). All hidden variables are non-linear functions of linear combinations of the $x$-variables. In the same way, the output variable $y$ is a non-linear function of linear combinations of the hidden variables. In this way the neural network function becomes a non-linear function of the $x$-variables. The non-linear functions in the hidden and the output layers can, in principle, be chosen as any continuous function, but they are most frequently taken to be sigmoid ones. The parameters in the network are the coefficients of the input $x$-variables and the coefficients of the hidden variables, and these can be found by, for instance, LS optimisation. The number of variables in the hidden layer (or hidden layers) can be found by testing different possibilities and selecting the best. Neural networks have become very popular, and used with caution they can be very useful in, for instance, NIR spectroscopy. Special attention should be given to validation of neural networks, because they are very flexible functions and therefore sensitive to overfitting. To overcome this problem it has been proposed to use principal components as input instead of the original variables themselves [see, for example, Næs *et al.* (1993) and Borggaard and Thodberg (1992)]. A drawback with neural networks is that they are still more difficult to use for interpretation of the data than some of the other methods discussed in this book.

### 9.3.8 Deleting non-linear variables

In some cases there may be some $x$-variables that have a more complex relationship with $y$ than others. For instance, Brown (1992) presented an application in NIR spectroscopy where deleting some of the wavelengths in the spectrum gave a great improvement in prediction ability using the PLS method. One possible explanation of this is an improvement in linearity: the non-linear wavelengths are deleted and only the ones that fit best into a linear model are used. See Chapter 7 for an explanation of how the most useful variables/wavelengths can be selected.

**9.3** An overview of different strategies for handling non-linearity problems

# 10 Scatter correction of spectroscopic data

This chapter is about mathematical techniques useful for reducing effects such as differences in sample thickness and light scatter in diffuse spectroscopy (for instance NIR, see Appendix B). The idea behind the techniques is that for calibration purposes, spectra should contain as little irrelevant information as possible. This is important both for the prediction performance and for interpretability of the spectra [see, for example, Isaksson and Næs (1988)].

**AA**
LS: least squares
MSC: multiplicative scatter correction
NIR: near infrared
OS: optimised scaling
OSC: orthogonal signal correction
PCA: principal component analysis
PLC-MC: path length correction with chemical modelling
PLS: partial least squares
PMSC: piecewise multiplicative scatter correction
SNV: standard normal variate

## 10.1 What is light scatter?

In diffuse reflectance/transmittance spectroscopy, light will be reflected and transmitted when the refractive index changes. Typically, this happens when the light meets a particle in a powder or a droplet surface in an emulsion. The scattering of light will then be a function of two properties:
- the number of light and surface interactions, depending on, for instance, the particles' or droplets' size and shape;
- the actual differences in refractive indices, implying also that the surroundings will influence the light's interaction with an analyte.

There will always be more light scatter or total reflection with small particles or droplets than with larger ones. There will also be more scatter the more different the refractive indices are. This is analogous to human vision; e.g. dry snow appears whiter than ice or wet snow.

The Kubelka–Munck (1931) theory states that when light penetrates a homogenous scattering material, it will either be absorbed or scattered. By analysing the light flux in the penetration direction in addition to the flux in the opposite direction, and by applying mathematical techniques, the so-called Kubelka–Munck function was developed:

$$K/S = (1 - R)^2/2R \qquad (10.1)$$

where $K$ is the true absorbance, $S$ is scatter and $R$ is the reflected light. Formally, $R$ is defined as $R = I/I_0$ where $I$ is the intensity of the reflected light and $I_0$ is intensity of incident light. According to this theory, the light scatter effect is multiplicative. This means that if the spectrum is first properly transformed to the Kubelka–Munck scale, a difference in scatter between two "equal" samples can be compensated for by multiplying the measurement at each wavelength of one of the samples by the same constant. A similar multiplicative effect can sometimes also be seen for another much used transform of reflectance data, namely the so-called absorbance $A = \log(1/R)$ [see, for example, Geladi *et al.* (1985)].

In diffuse NIR spectroscopy it also sometimes happens that there are clear indications of an additive scatter component [Geladi *et al.* (1985)]. One reason for such an effect may be the following: the theoretical models (e.g. Kubelka–Munck and Beer–Lambert) assume that all or a constant part of the reflected light is detected. But if the instrument is constructed in such a way that only a fraction ($1/c$) of the reflected light is detected for a particular sample, we get

---

Kubelka–Munck and Beer–Lambert are theoretical models for spectroscopy. See also Appendix B.

$$I_{detected} = 1/c \times I_{reflected}$$
$$\begin{aligned}A_{detected} &= -\log(R_{detected}) = -\log(I_{detected}/I_0) \\ &= \log c + \log(I_0/I_{reflected}) = c' + A\end{aligned} \quad (10.2)$$

If $c' = \log(c)$ is sample dependent, this will cause an additive baseline difference between the samples, i.e. an additive effect in the absorbance values.

## 10.2 Derivatives

An easy to perform and often used method to reduce scatter effects for continuous spectra is to use derivatives.

The solid line in Figure 10.1 shows a transmission NIR spectrum of a sample of whole wheat, measured at 2 nm intervals over the range 850 to 1048 nm. Also shown are two derivative spectra, the first derivative as a dashed line, the second as a dotted line. The vertical scale on this plot is arbitrary, the curves have been rescaled so that they all fit on the same plot. The zero is, however, in the correct place for the two derivative spectra. The first derivative spectrum is the slope at each point of the original spectrum. It has peaks where the original has maximum slope, and crosses zero at peaks in the original. The second derivative is the slope of the first derivative. It is more similar to the original in some ways, having peaks in roughly the same places, although they are inverted in direction. It is a measure of the curvature in the original spectrum at each point.

Taking the first derivative removes an additive baseline. A spectrum parallel to the one in Figure 10.1 but shifted upwards or downwards would have the same first derivative spectrum, because the slope would be the same everywhere. Taking a second derivative removes a linear baseline. A straight line added to the original spectrum becomes a constant

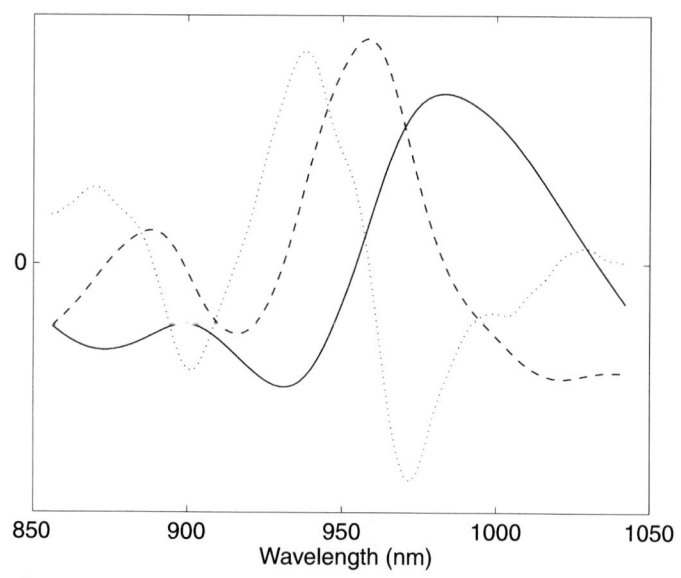

**Figure 10.1.** A transmission spectrum of a sample of whole wheat, measured at 2 nm intervals from 850 to 1048 nm (solid line), together with its first (dashed line) and second (dotted line) derivatives.

shift in the first derivative, because the straight line has a constant slope, and this is removed in the second derivative. An additive linear baseline shift is not the same as a multiplicative scatter effect, but it may look very similar in practice.

What the discussion so far has ignored is the fact that the measured spectrum is not a continuous mathematical curve, but a series of measurements at equally-spaced discrete points. The spectrum in Figure 10.1 is plotted again in Figure 10.2, showing the individual points instead of joining them up in the usual way. An obvious way to calculate derivatives with such data is to use differences between the values at adjacent points. Thus the first derivative at wavelength $w$ might be calculated as $y_w - y_{w-1}$, where $y_w$ is the measured spectrum at wavelength number $w$ in the sequence. Strictly, one should divide by the wavelength gap (in nm or other appropriate units) between

**10.2 Derivatives**

# Scatter correction of spectroscopic data

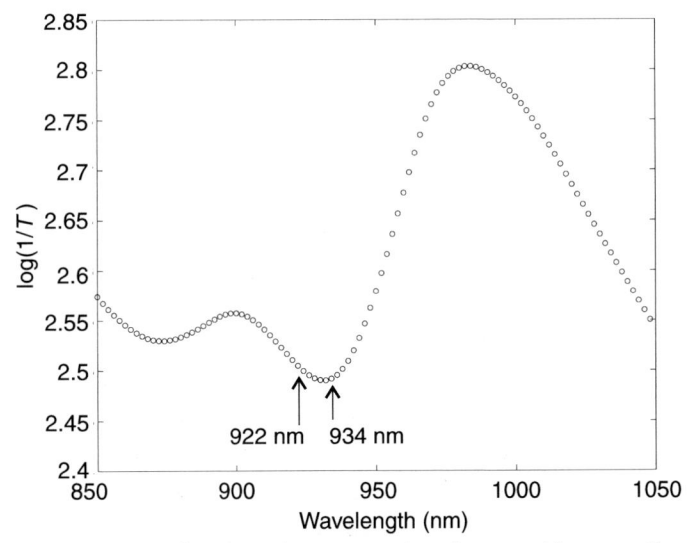

**Figure 10.2.** The actual measured points making up the spectrum in Figure 10.1. The arrows indicate the window that is enlarged in Figure 10.3.

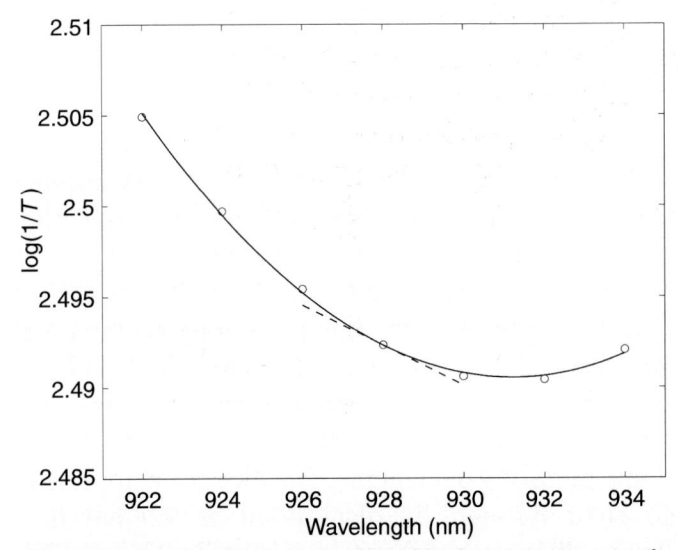

**Figure 10.3.** A quadratic curve fitted by least squares to the seven data points from 922 to 934 nm. The dashed line is the tangent to the fitted curve at 928 nm.

## 10.2 Derivatives

points to get a true slope, but this only affects the scaling of the derivative and so it is not usually done. The second derivative would then be the difference of two adjacent first derivatives, resulting in $y_{w-1} - 2y_w + y_{w+1}$, the second difference formula.

The problem with this simple approach is that differencing reduces the signal and increases the noise at one and the same time. Unless the measurements are error-free the result of these calculations will be a noisy derivative spectrum. In practice it is almost always necessary to incorporate some smoothing into the calculation. One way to do this is to use simple first or second differences as described above, but base these on averages over several points. Thus for a second derivative one can select three narrow intervals, or windows, average the points in each window, and apply the second-difference formula to these averages. This approach is widely used for NIR spectra because it is provided in popular software.

A somewhat more elegant approach, described by Savitzky and Golay (1964), was used to produce the derivative spectra in Figure 10.1. Figure 10.3 shows the estimation of the derivatives at 928 nm. We take a narrow window centred at the wavelength of interest and fit a low-order polynomial to the data points in the window using least squares. Here a seven-point window, 928 nm and three wavelengths on either side, has been used, and a quadratic curve fitted. The fitted curve is shown as the solid line in Figure 10.3. It does not go through the points (it only has three parameters and there are seven points so it cannot generally achieve this) but is pretty close to all of them, because the spectrum is nearly quadratic over such a small range of wavelengths. The fact that it appears to go right through the data point at 928 nm is a fluke—there is no reason why it has to in general. This quadratic is a continuous curve, with equation

$$y = \hat{a} + \hat{b}x + \hat{c}x^2 \qquad (10.3)$$

where $x$ is wavelength, $y$ is the spectral absorbance and $\hat{a}$, $\hat{b}$ and $\hat{c}$ are the parameters estimated by the least squares fit. It has a slope

$$dy/dx = \hat{b} + 2\hat{c}x \qquad (10.4)$$

that can be evaluated for any $x$, and a second derivative

$$d^2y/dx^2 = 2\hat{c} \qquad (10.5)$$

that is constant for all $x$. The first derivative is easy to picture, it is the slope of a tangent to the curve. The tangent at 928 nm is shown by the dashed line in Figure 10.3. The second derivative measures the curvature, and is harder to illustrate. It is the rate of change of the slope as we move along the curve.

The idea behind the Savitzky–Golay approach is to use the first and second derivative of the fitted curve at 928 nm to estimate the first and second derivative of the underlying spectrum at 928 nm. If the spectrum is well-fitted locally by a quadratic, this procedure will give good estimates of the "true" derivatives.

To estimate the derivatives at 930 nm, the next point along, we slide the window along so that it now runs from 924 nm to 936 nm, i.e. still has width seven points but is now centred on 930 nm, and repeat the entire process. Thus the seven-point fit is only used to produce derivatives at one point in the centre of the window, not at seven.

What the description above fails to make clear is just how simple the computations for this can be made. When we fit a polynomial (for example, a quadratic) curve to $(x, y)$ data, the coefficients are linear combinations of the $y$s with weights that are functions of the $x$s (the wavelengths) and, as can be seen from the equations above, the derivatives are very simple

**10.2 Derivatives**

functions of the coefficients. With equally spaced wavelengths it is fairly easy to work out that the estimated first derivative at 928 nm is

$$(-3y_{922} - 2y_{924} - y_{926} + 0y_{928} + y_{930} + 2y_{932} + 3y_{934})/28$$

and the estimated second derivative at 928 nm is

$$(5y_{922} + 0y_{924} - 3y_{926} - 4y_{928} - 3y_{930} + 0y_{932} + 5y_{934})/42$$

Here $y_{922}$ denotes the spectral measurement at 922 nm and so on. To get the derivatives at the next point you use the same weights, (–3 –2 –1 0 1 2 3)/28 for the first and (5 0 –3 –4 –3 0 5)/42 for the second derivative, but apply them to a sequence of $y$s starting at $y_{924}$ and ending at $y_{936}$. The weights only need to be computed once (or looked up somewhere, for example in the original paper) and you just run them along the spectrum doing a simple weighted average calculation to get the derivatives. This type of calculation is called linear filtering by engineers.

As with the simple differences, these weights get the scaling of the derivatives wrong. They are based on an assumed gap of one unit between consecutive $x$s, whereas the interval here is 2 nm. The actual first derivative should be half what has been computed. However, nobody ever seems to bother about this, since the vertical scale on a derivative spectrum is rarely looked at.

The somewhat arbitrary choice of seven points for the window needs some discussion. The choice of window size (or filter length) involves a trade-off between noise reduction or smoothing (for which we want as wide a window as possible) and distortion of the curve, which will happen if the window is too wide. The effect of using too wide a window is to round off the peaks and troughs—you can see this just starting to happen in the fit at 930 and 932 nm in Figure 10.3. The problem with a narrow window is that the noise in the original spectrum is inflated by the de-

> You need to read the original paper, but there were some errors in the weights. These were corrected by Steinier, Termonia and Deltour (1972).

rivative calculation. If you use a three-point window, in which case the Savitzky–Golay filter for the second derivative has weights of (1 –2 1) and corresponds to the simple second difference, for this spectrum the result looks quite noisy in places. One way to find a suitable width would be to start with three points and increase the window width until the derivative spectra are not visibly noisy.

There is another choice to be made. A quadratic curve was also used here, but a cubic, i.e. one with an $x^3$ term, is sometimes used. Such a curve is more flexible, but needs a wider window to achieve the same amount of noise reduction, so it is not at all clear whether it is to be preferred or not. To decide what works best one needs to study the effect of various choices on noise and distortion for one's particular spectra, which can be time-consuming. Using the default settings in instrument manufacturer's software is an attractive option in these circumstances.

These filters have been described as though they are only used for calculating derivatives, but the same polynomial fit can also be used for smoothing. We just replace the observed value at the midpoint of the window by the fitted value on the curve. This can be done by applying yet another filter, with weights (–4 6 12 14 12 6 –4)/42 in this case, to the data. For the spectrum in the example it is not clear that any smoothing is desirable (perhaps because the instrument software has already smoothed it), but for noisier ones it might be. There is more discussion of smoothing in Section 8.2 on Fourier methods.

The Savitzky–Golay approach is more elegant than averaging over intervals, but are the results any better? If the two methods are compared on artificial examples it is clear that Savitzky–Golay gives a less distorted estimate of the "true" derivative. If, however, one thinks of the derivative calculations as a spectral pre-treatment, rather than an attempt to esti-

**10.2 Derivatives**

mate some true derivative, it does not necessarily follow that a more accurate derivative is the route to a better calibration. It is not at all clear which method is to be preferred for this purpose, and the answer almost certainly varies from case to case, although the differences are unlikely to be important.

A possible disadvantage of using derivatives is that they change the form of the original spectrum. Second derivatives have the advantage that peaks appear at similar locations in the second derivative and the original spectrum. However, the second derivative spectrum in Figure 10.1 has more features than the original, and this is generally true. When the original spectrum is as featureless as this one, this may be seen as an advantage. When the original is already complex, any increase in the difficulty of interpretation is unwelcome.

## 10.3 Multiplicative scatter correction (MSC)

In a later paper [Martens and Stark (1991)], Martens and Stark prefer to call it multiplicative *signal* correction as this does not make any assumptions about the required correction of the signal.

The MSC method [see, for example, Martens *et. al*. (1983) and Geladi *et al*. (1985)], originally developed for NIR data, is most easily motivated by looking at plots of individual spectral values versus average spectral readings as in Figure 10.4. One can see that the spectral values for each of the samples lie quite close to a straight regression line. Such different regression lines were interpreted by Geladi *et al*. (1985) as differences due to scatter effects, while the deviations from the regression lines were interpreted as the chemical information in the spectra.

As can also be seen, the regression lines for the different samples need an intercept in order to be fitted properly. These empirical findings indicate that the "scatter effect" for these samples is not only multiplicative, but can also have a significant additive com-

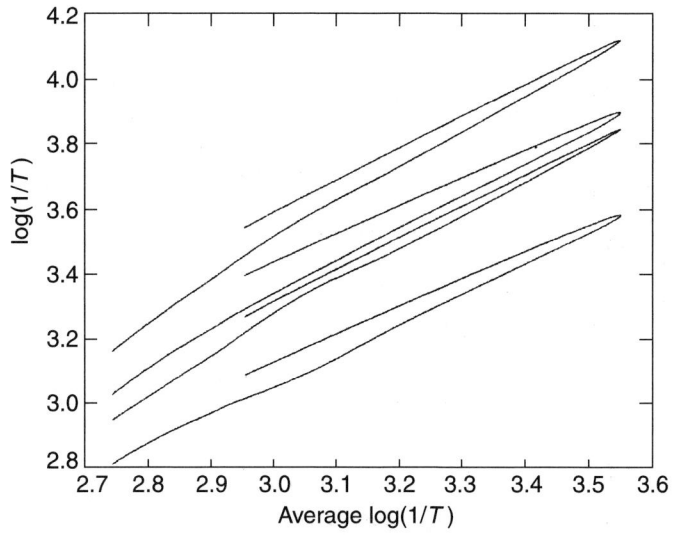

**Figure 10.4.** Plot of average versus individual NIR spectral values for four spectra of meat (absorbance units).

ponent. The MSC model for each individual spectrum is

$$x_{ik} = a_i + b_i \bar{x}_k + e_{ik} \quad (i = 1, \ldots, N; k = 1, \ldots, K) \quad (10.6)$$

where $i$ is the sample number and $k$ is the wavelength number. The constant $a_i$ represents the additive effect while $b_i$ represents the multiplicative effect for sample $i$. The mean

$$\bar{x}_k = \frac{1}{N} \sum_{i=1}^{N} x_{ik}$$

is the average over samples at the $k$th wavelength.

The $a_i$ and $b_i$ coefficients are unknown and must be estimated individually for each sample using all or a subset of the $K$ spectral measurements (see below). The "error" $e_{ik}$ in the model corresponds to all other effects in the spectrum that cannot be modelled by an additive and multiplicative constant. Once the con-

**10.3** Multiplicative scatter correction (MSC)

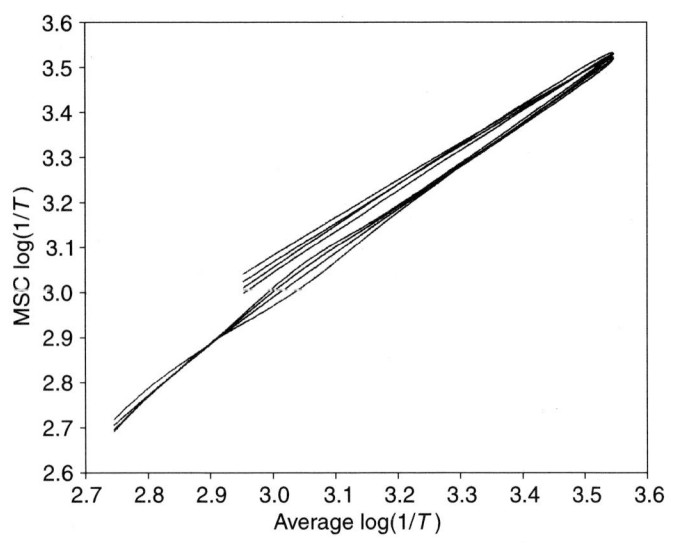

**Figure 10.5.** The effect of MSC. The same four spectra that were plotted in Figure 10.4 are now scatter corrected.

stants $a_i$ and $b_i$ are estimated by LS, they are used in the MSC transform which subtracts $\hat{a}_i$ from $x_{ik}$ and divides the result by $\hat{b}_i$:

$$x_{ik}^* = (x_{ik} - \hat{a}_i)/\hat{b}_i \qquad (10.7)$$

The effect of this transform on the data in Figure 10.4 is illustrated in Figure 10.5. Most of the variation among the spectra is eliminated, i.e. the dominating additive and multiplicative effects have been removed. The corresponding spectra before and after scatter correction are presented in Figure 10.6 and Figure 10.7.

The same technique can be used if there are only multiplicative effects in the "model". One simply eliminates the parameter $a_i$ from equation (10.6), estimates $b_i$ by standard LS and divides the $x_{ik}$ by $\hat{b}_i$. In the same way, $b_i$ can also be removed from the model. In this case, the additive constant $a_i$ is estimated by the average of $x_{ik}$ over all wavelengths in the spectrum. In general, we recommend the use of model (10.6).

# Scatter correction of spectroscopic data

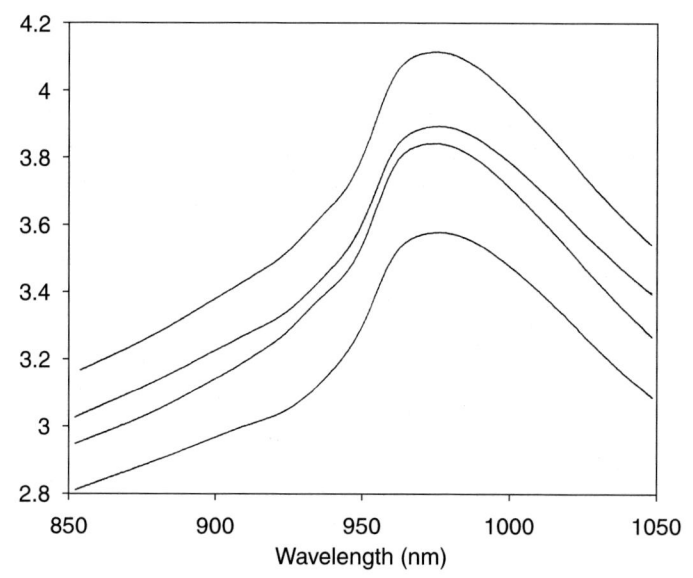

**Figure 10.6.** NIR spectra before scatter correction.

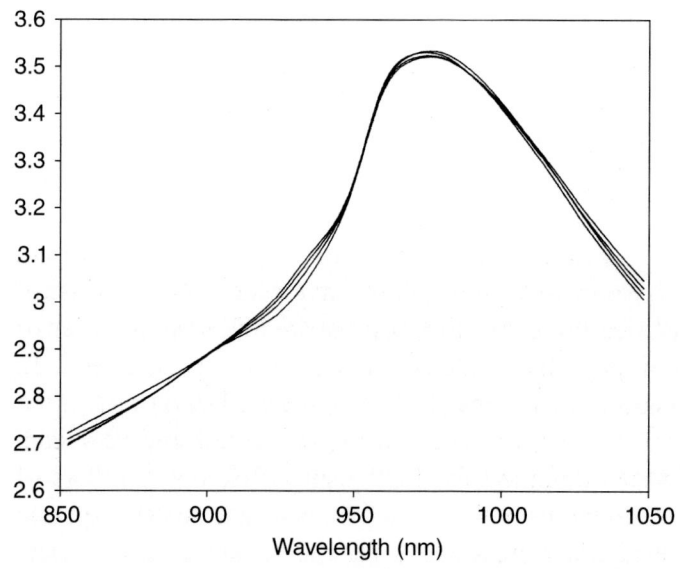

**Figure 10.7.** The NIR spectra in Figure 10.6 after scatter correction.

MSC works primarily for cases where the scatter effect is the dominating source of variability. This is very typical in, for instance, many applications of dif-

**10.3** Multiplicative scatter correction (MSC)

fuse NIR spectroscopy. If this is not the case, the chemical information in the spectrum, which corresponds to the "error" in equation (10.6), will have too much influence on the slope and intercept of the regression line. The MSC transform will then usually be too dependent on chemical information and may remove some of it.

In order to use MSC safely, the sum of all the constituents that absorb light in the actual region should preferably be equal to a constant, for instance 100%. If this is not the case, it may become impossible to distinguish between the effect of the sum of the constituents and the scatter. The reason for this is that both may have a similar multiplicative effect on the spectrum.

MSC can be applied to all the wavelengths or only a subset of them. If there is a spectral region which is less dependent on chemical information than other regions, the $a$ and $b$ values can be computed by using only this spectral range. The MSC transform in (10.7) is, of course, applied to all the wavelengths. In this way, the constants $\hat{a}_i$ and $\hat{b}_i$ become as independent of the chemical information as possible.

From a calibration point of view, two important consequences of the MSC have been observed; it simplifies the calibration model (reduces the number of components needed) and may improve linearity. In Isaksson and Næs (1988) the squared sum of the fitted residuals $\hat{f}$ for both scatter-corrected and raw data were computed for PCR using different numbers of components in the model. It was clear that the squared residuals were smaller for the scatter-corrected data than for the corresponding model before scatter correction. The prediction results using the calibration equations with and without MSC were also computed. The number of principal components required to obtain good results tended to be smaller for the scatter-corrected data.

**10.3** Multiplicative scatter correction (MSC)

The MSC method has been used in a large number of NIR applications. In many of these cases it has given improved prediction results. The relation between MSC and some other and similar transforms is clarified in Helland *et al.* (1995). One of the results obtained was that the MSC is both conceptually and empirically strongly related to the standard normal variate method [SNV, Barnes *et al.* (1989)], described in Section 10.8.

## 10.4 Piecewise multiplicative scatter correction (PMSC)

This MSC method can easily be extended to a more flexible and general correction procedure in which each wavelength is corrected using individual additive and multiplicative terms. One way of doing this is by using the same idea as for MSC over a limited wavelength region. Since this is analogous to MSC, the method is called piecewise MSC [PMSC, Isaksson and Kowalski (1993)]. For each spectral value $x_{ik}$, constants $a_{ik}$ and $b_{ik}$ are estimated by regressing spectral values on averages in a wavelength neighbourhood of $x_{ik}$ only. The correction of $x_{ik}$ is done as for MSC.

In practice, a window-size optimisation is needed. This can be done by, for instance, cross-validation in the same way as optimisation of the number of components in PLS and PCR regression (see Chapter 13). Results in Isaksson and Kowalski (1993) indicate that this type of pre-processing of data can in some cases lead to substantial improvements in prediction ability compared to non-corrected and MSC corrected data.

In Figure 10.8 is given an illustration of the effect of PMSC compared to MSC on NIR transmission spectra of meat. The PMSC, using a 20 nm moving window, removes more variability than does the

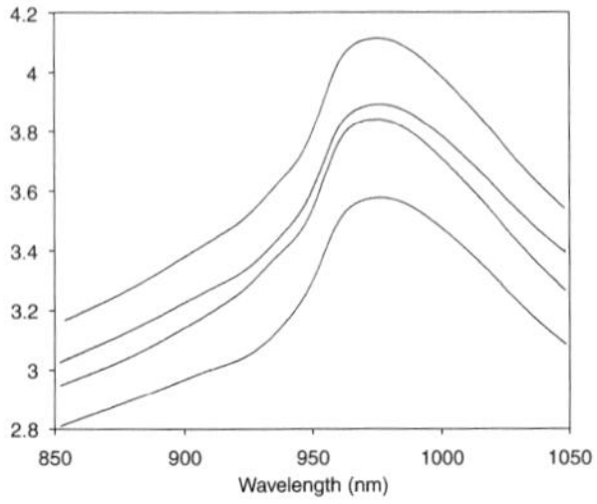

**Figure 10.6.** NIR spectra before scatter correction.

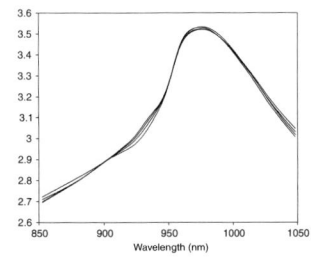

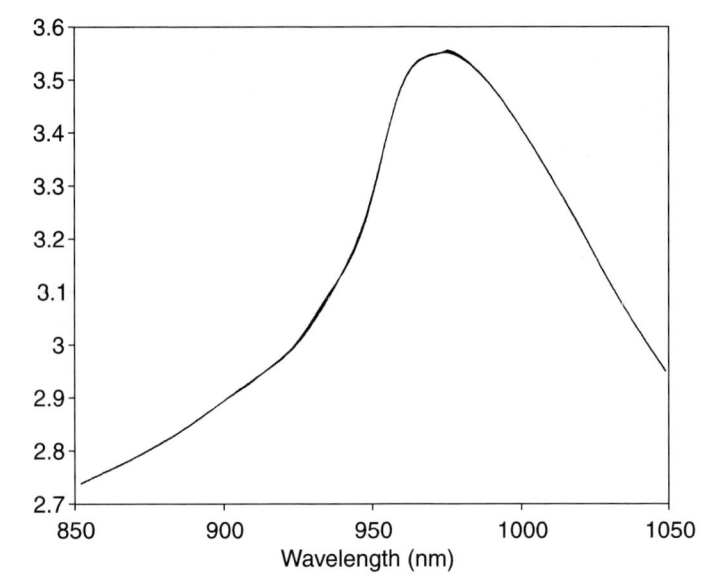

**Figure 10.8.** The spectra in Figure 10.6 after PMSC.

**Figure 10.7.** The NIR spectra in Figure 10.6 after scatter correction.

MSC. Visually, there is very little information left in the spectra after PMSC, but for this meat application, prediction results based on PMSC data (68 samples for calibration and 32 for testing) gave substantially better results than results obtained for the non-corrected and MSC-corrected data. Compared to MSC data, PMSC gave 22%, 19% and 31% improvement for prediction of protein, fat and water, respectively.

## 10.5 Path length correction method (PLC-MC)

Another scatter correction technique was proposed in Miller and Næs (1990) for the special situation that all the samples in the calibration set can be controlled to have the same scatter or path length level. In addition, it was assumed that the scatter/path length effect was totally multiplicative. A new spectrum with possibly different path length level must then be corrected using a multiplicative model before it is predicted. The method was named "path length correction with chemical modelling" (PLC-MC).

The first step of the transform is to model the calibration data using PCA. A new spectrum, $\mathbf{x}_p$, with a possibly different path length level is then modelled using the equation

$$a\mathbf{x}_p = \bar{\mathbf{x}} + \hat{\mathbf{P}}\mathbf{t}_p + \mathbf{e} \quad (10.8)$$

with $\mathbf{t}_p$ representing the principal component scores of the spectrum after the transform. As before $\hat{\mathbf{P}}$ is the matrix of PCA loadings and $\bar{\mathbf{x}}$ is the average spectrum. The model is then rearranged by multiplying each side by $a^* = 1/a$ giving

$$\mathbf{x}_p = a^*\bar{\mathbf{x}} + \hat{\mathbf{P}}\mathbf{t}_p^* + \mathbf{e}^* \quad (10.9)$$

In particular, this method can be useful for in-line NIR spectroscopy. In such cases, the probe needs maintenance and cleaning and it may be difficult to put it back in exactly the same position. This may result in effects like those explained here. See Sahni et al. (2001).

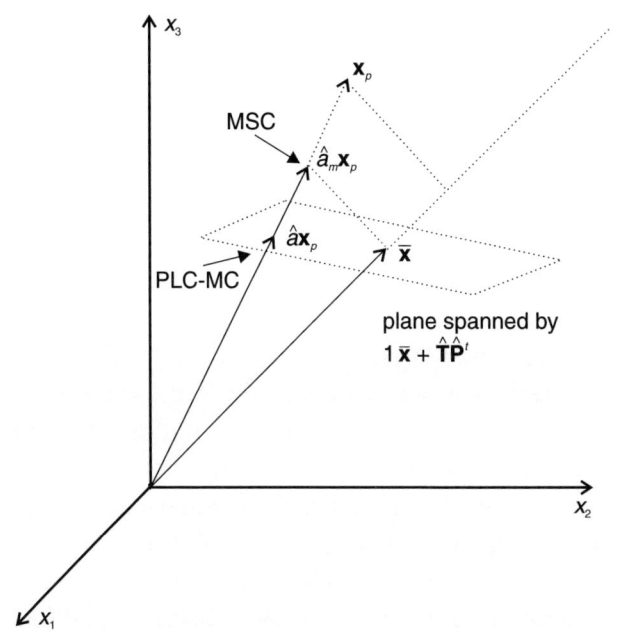

Figure 10.9. The PLC-MC method illustrated graphically. A prediction spectrum ($\mathbf{x}_p$), the prediction spectrum corrected by the MSC method ($\hat{a}_m\mathbf{x}_p$) and the prediction spectrum corrected by the PLC-MC method ($\hat{a}\mathbf{x}_p$). Reproduced with permission from C. Miller and T. Næs, *Appl. Spectrosc.* 44, 896 (1990) © Society for Applied Spectroscopy.

**10.5 Path length correction method (PLC-MC)**

The coefficients $a^*$ and $t_p^*$ are then determined by LS regression. The path length corrected spectrum is then found by dividing $\mathbf{x}_p$ by $\hat{a}^*$. See Figure 10.9 for an illustration.

The method performed well in an example of NIR analysis of a set of polymer samples where the thickness of the calibration samples could be controlled. The method was modified and extended in Miller *et al.* (1993) to handle certain types of collinearities observed when fitting the model (10.9) by LS.

## 10.6 Orthogonal signal correction (OSC)

Orthogonal signal correction (OSC) is a method developed for reducing light scatter effects, and indeed more general types of interferences, whilst only removing those effects that have zero correlation with the reference value $y$. The idea is that all information in the spectrum related to $y$ should remain rather than be removed.

The first step of the algorithm is to compute loading weights, $\hat{\mathbf{w}}$, such that the score vector $\hat{\mathbf{t}} = \mathbf{X}\hat{\mathbf{w}}$ describes as much as possible of the variance of $\mathbf{X}$ under the constraint that it is uncorrelated with $\mathbf{y}$. After having obtained such a component, its effect is subtracted before a new factor of the same type is computed. The procedure can continue, but usually a very small number of "correction" components is needed. The residuals after the use of OSC are used for regular PLS/PCR modelling. We refer to Wold *et al.* (1998) for further details.

Some empirical results are promising [Wold *et al.* (1998)] while others are less so [Fearn (2000)]. More research is needed before final conclusions about the merits of the method can be drawn.

## 10.7 Optimised scaling (OS)

Optimised scaling (OS) models the multiplicative scatter effect and the regression coefficients in one single equation

$$\mathbf{Y} = \mathbf{MXB} + \mathbf{F} \qquad (10.10)$$

Here $\mathbf{Y}$ is the matrix of centred reference values ($J$ constituents), $\mathbf{M}$ is a diagonal matrix with multiplicative scatter terms on the diagonal, $\mathbf{X}$ is the centred spectral matrix, $\mathbf{B}$ is a matrix of regression coefficients and $\mathbf{F}$ contains the residuals.

For mathematical simplicity, $\mathbf{M}$ is inverted and placed on the "Y-side" of the equation. The scatter parameters in $\mathbf{M}$ and regression coefficients $\mathbf{B}$ are then found using a simultaneous fit [see Karstang and Manne (1992) for details].

To avoid the null solution, $\mathbf{M} = 0$ and $\mathbf{B} = 0$, one sample must be selected as a reference and the multiplicative term for this sample is usually set to 1. Other choices were discussed in Isaksson *et al.* (1993).

From a mathematical point of view the OS approach is very attractive and the algorithm is reasonably simple. One of its possible advantages over MSC is the simultaneous estimation of the scatter effects and regression coefficients. This can possibly make the estimated scatter effects less dependent on chemical variation in the spectra. Note, however, that optimised scaling will not correct for additive spectral variation as do both MSC and PMSC. This "problem" can be overcome if the calibration is performed on derivative spectra. Some empirical results indicate that the OS method has potential for improving the prediction results as compared to the MSC method [Karstang and Manne (1992)], but other results are inconclusive [Isaksson *et al.* (1993)].

Note that data compression by, for instance PCA, is needed here also.

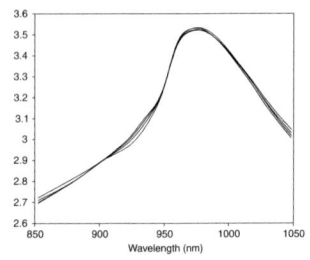

**Figure 10.7.** The NIR spectra in Figure 10.6 after scatter correction.

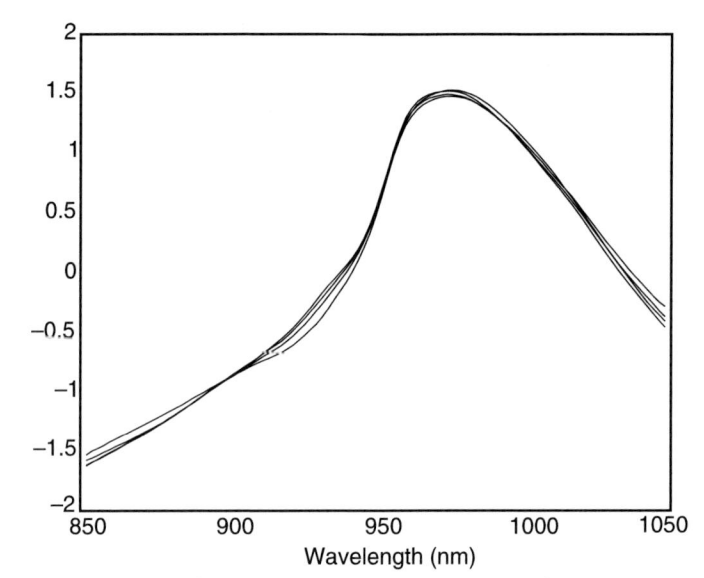

**Figure 10.10.** The NIR spectra in Figure 10.6 after SNV.

## 10.8 Standard normal variate method (SNV)

This method, described by Barnes *et al.* (1989), centres and scales individual spectra, having an effect very much like that of MSC. If, as before, $x_{ik}$ is the spectral measurement at the $k$th wavelength for the $i$th sample, the transformed value is

$$x_{ik}^* = (x_{ik} - m_i) / s_i \qquad (10.11)$$

where $m_i$ is the mean of the $K$ spectral measurements for sample $i$ and $s_i$ is the standard deviation of the same $K$ measurements. Figure 10.10 shows the spectra in Figure 10.6 after the application of SNV. The effect of the SNV is that, on the vertical scale, each spectrum is centered on zero and varies roughly from −2 to +2. Apart from the different scaling, the result is similar to that of MSC, as can be seen by comparing Figures 10.7 and 10.10. The main practical difference is that SNV standardises each spectrum using only the data from that spectrum; it does not use the mean

> This multiplicity of methods (and this text is not a complete list) will appear confusing to the novice. Fortunately, most software packages will offer only a limited choice, probably derivatives and MSC or SNV. These will be adequate for a large proportion of problems which contain data affected by scatter.

spectrum of any set. The choice between the two is largely a matter of taste and the availability of software [see also Helland *et al.* (1995)].

**10.8** Standard normal variate method (SNV)

# 11 The idea behind an algorithm for locally weighted regression

In Chapter 9 we discussed alternative strategies for solving non-linearity problems. We pointed at non-linear calibration as one useful approach and some of the actual possibilities were mentioned. In particular, we discussed briefly the concept of locally weighted regression (LWR) and indicated some of its advantages. The present chapter will follow up these ideas and present the computational strategy for LWR and also give some more information about the performance of the method.

**AA**
CARNAC: Comparison Analysis using restructured Near infrared And Constituent data
LS: least squares
LWR: locally weighted regression
PCA: principal component analysis
PCR: principal component regression
PLS: partial least squares
*RMSEP*: root mean square error of prediction
WLS: weighted least squares

## 11.1 The LWR method

Most regression surfaces can be approximated locally by linear models. LWR utilises this idea for each new prediction sample by first seeking samples which are close to it in spectral space and then using these local samples to fit a local linear regression equation [see, for example, Cleveland and Devlin (1988) and Næs *et al.* (1990)].

In many practical applications in chemistry, the calibration **X**-matrix is often highly collinear. Therefore, the raw *x*-data must first be compressed, by principal components analysis for example, before LWR can be applied.

The algorithm for LWR goes as follows:
1. Decide how many principal components (*A*) and how many samples (*C*) are to be used in each local calibration set.

2. Use PCA on the full calibration set. As usual we let $\hat{\mathbf{T}}$ be the score matrix and $\hat{\mathbf{P}}$ be the loading matrix corresponding to the first $A$ components.

For each prediction sample:

3. Compute the scores $\hat{\mathbf{t}}' = \mathbf{x}'\hat{\mathbf{P}}$, where $\mathbf{x}$ is the prediction spectrum centred by the same constants as used to centre the calibration matrix $\mathbf{X}$.
4. Find the $C$ calibration samples that are closest to the prediction sample. Closeness is defined by using a distance measure in the $A$-dimensional principal component score space. Distances between the $A$-dimensional score vector $\hat{\mathbf{t}}$ and all the rows of $\hat{\mathbf{T}} = \mathbf{X}\hat{\mathbf{P}}$ are calculated and used to find the $C$ samples that are closest. Figure 11.1 illustrates the idea behind this, using a single $x$-variable.
5. The least squares (LS) criterion is used to estimate the unknown parameters in the linear regression model

$$\mathbf{y}_C = \mathbf{1}q_0 + \hat{\mathbf{T}}_C\mathbf{q} \qquad (11.1)$$

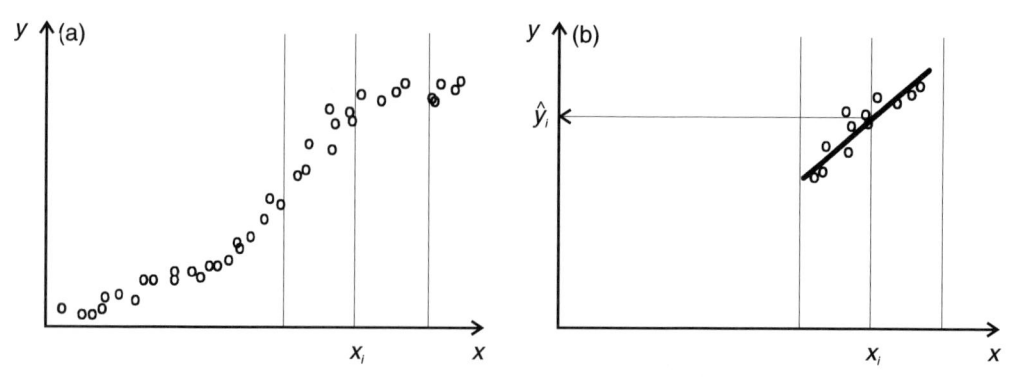

Figure 11.1. The LWR principle. (a) For each new prediction sample (the $x_i$-value) a number of local samples are identified and used to fit a local linear model. (b) The $y$-value of the new sample is predicted using this local model.

Here $\mathbf{y}_C$ is the vector of y-values for the $C$ local samples, $\mathbf{q}$ is the corresponding vector of regression coefficients for the $A$ scores, $\hat{\mathbf{T}}_C$ denotes the matrix of scores for the $C$ samples closest to the prediction sample, $\mathbf{1}$ is a vector of 1s and $q_0$ is the intercept term. If a predefined weighting function is available, weighted least squares [WLS, see Næs *et al.* (1990)] can be used instead of regular LS.

> Weighting allows some observations to be considered more important than others.

6. Predict the y-value for the prediction sample using the estimated regression coefficients,

$$\hat{y} = \hat{q}_0 + \hat{\mathbf{t}}'\hat{\mathbf{q}} \qquad (11.2)$$

Note that this whole procedure is described for local *linear* approximations (see equation 11.1). Local quadratic functions can be used instead if a more flexible local modelling is needed [see, for example, Næs *et al.* (1990)].

The LWR method needs a new calibration for each prediction object. At first glance, this may look time-consuming and cumbersome. In practice, however, one usually needs only a small number of samples in each local set and also a quite small number of principal components, so from a computational point of view this will usually represent no problem.

Note that the LWR method is based on already well-known concepts such as PCA, linear regression and a simple distance measure to define closeness. This implies that all plots and standard linear regression diagnostic tools, such as leverage and influence, are directly available (see Chapter 14).

It should also be noted that there are a number of tuning parameters that have to be determined for the LWR. One has to decide what kind of distance measure to use, which weight function to apply in the WLS regression (see Appendix A) and how many

**11.1 The LWR method**

components, $A$, and samples, $C$, to use in each local regression.

## 11.2 Determining the number of components and the number of local samples

*C: the number of local samples*
*A: the number of components*

The number of components ($A$) and the number of local samples ($C$) which give the best results can be determined in each particular case in the same way as the number of components is determined for PCR or PLS. In practice, cross-validation and prediction testing are good alternatives (see Chapter 13). An illustration of this is given in Figure 11.2. This is based on prediction testing in an application to NIR measurements of meat. The LWR method is used and tested for a number of different values of $C$ and $A$. As can be seen, the smallest value of *RMSEP* is obtained by using a value of $C$ equal to 35 and an $A$ equal to 3. These values are good candidates to be used for future predictions. As for selection of components in PCR/PLS, the $A$ value should also be as small as possible here.

Note that each local regression in the algorithm is based on the same principal components as computed for the whole calibration set. An interesting

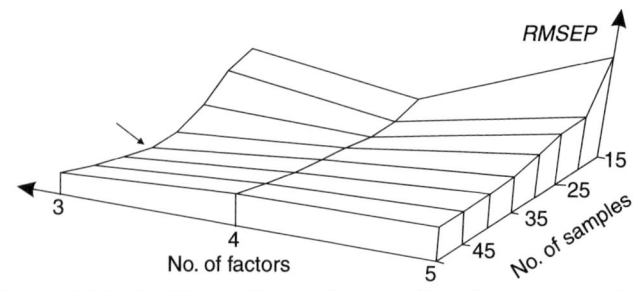

**Figure 11.2.** An illustration of how to find the best possible *C* and *A* values using prediction testing. *RMSEP* is computed for a number of (*C*, *A*) combinations and the best/lowest value is selected. Reproduced with permission from T. Næs and T. Isaksson, *Appl. Spectrosc.* 46, 34 (1992) © Society for Applied Spectroscopy.

modification of this procedure was proposed in Aastveit and Marum (1993). They proposed to base each local estimation on a new set of principal components computed from the local set of samples. In this way more relevant principal components are expected. On the other hand they may become unstable. The results so far indicate a potential for improvement over the original algorithm, but it should be tested in more applications before a final conclusion can be made. Shenk *et al.* (1997) proposed to use PLS for each local calibration.

Note that since LWR is based on an idea of using only local samples in each regression equation, one will need "local information" over the whole region of interest. This is a strong argument for selecting calibration samples as evenly as possible. We refer to Chapter 15 for more information about how to select calibration samples in practice.

## 11.3 Distance measures

In step 4 of the algorithm, the problem is to find the $C$ samples that are "closest" to the prediction sample. The simplest way of doing this is to use the Euclidean distance, $d_1$, in the full spectral space defined by

$$d_1^2 = \sum_{k=1}^{K} (x_k - x_{pk})^2 \qquad (11.3)$$

As before $K$ is the number of wavelengths, $x_1$, ..., $x_K$ are the spectral measurements of a calibration sample and $x_{p1}$, ..., $x_{pK}$ are the corresponding values for the prediction sample. The Euclidean distance gives the same weight to all the directions in the spectral space.

An alternative is the Mahalanobis distance, which weights the principal component directions according to the variability along them. Since information in the principal components that correspond to

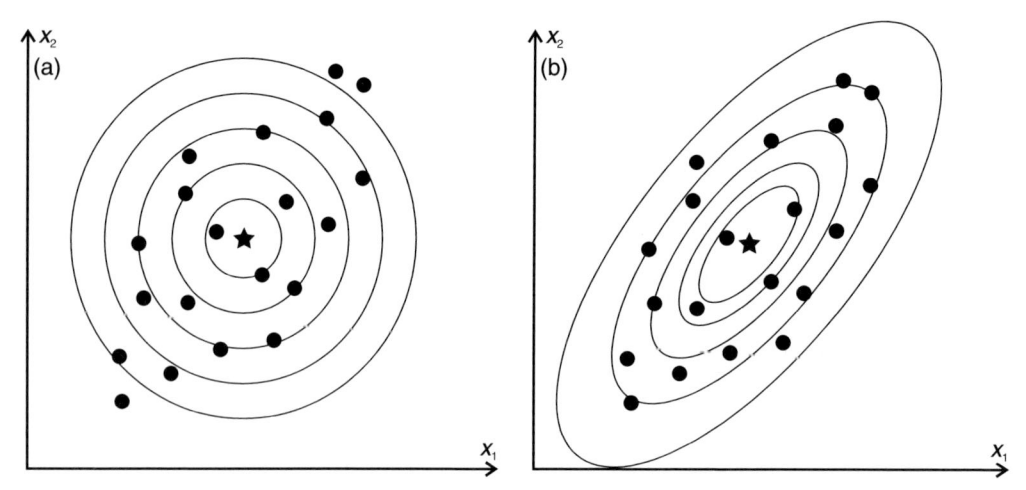

Figure 11.3. An illustration of different distance measures, (a) Euclidean and (b) Mahalanobis. The Euclidean distance is constant on circles around a point. The Mahalanobis distance is constant on ellipses following the general distribution of points.

the smaller eigenvalues are irrelevant in most NIR calibrations (see Chapter 5), such Mahalanobis distances should be truncated at $A$, the number of principal components used. The truncated Mahalanobis distance, $d_2$, between a prediction sample and a calibration sample can then be written as

$$d_2^2 = \sum_{a=1}^{A} (\hat{t}_a - \hat{t}_{pa})^2 / \hat{\lambda}_a \qquad (11.4)$$

Here $\hat{t}_a$ is the $a$th principal component score for a calibration sample and $\hat{t}_{pa}$ is the corresponding value for the prediction sample. As before, $A$ is the number of principal components used and $\hat{\lambda}_a$ is the eigenvalue of principal component $a$.

This distance measure is used in Næs et al. (1990). Figure 11.3 illustrates the difference between a Euclidean distance and a Mahalanobis distance.

A further modification is to let the influence of the different principal components depend on their prediction ability. This is done for the so-called modified Mahalanobis distance which was proposed in Næs and Isaksson [1992(b)]. Estimates of the prediction ability of the different components were obtained from root mean square error values (*RMSEP*, Chapter 13) from a linear PCR calibration. The modified distance, $d_3$, is defined by

$$d_3^2 = \sum_{a=1}^{A} [(\hat{t}_a - \hat{t}_{pa})^2 / \hat{\lambda}_a](RMSEP_{a-1}^2 - RMSEP_a^2) \quad (11.5)$$

where $RMSEP_a$ is the *RMSEP* obtained using $a$ components in the PCR model. Empirical results indicate that $d_3$ may be better in practice than $d_2$ [see, for example, Næs and Isaksson (1992b)] but more research is needed to evaluate their relative merits.

An illustration of how *RMSEP* values are used and how the different sample sets are selected is given in Figure 11.4.

Another modified distance measure based on the Mahalanobis distance was proposed in Wang *et al.* (1994) and called LWR2. They proposed to compute the distance as a weighted sum of a Mahalanobis distance and a distance between $y$ and $\hat{y}$. Since the $y$ is unknown at this stage, it must be estimated by, for instance, PCR.

The CARNAC method [Davies *et al.* (1988)] and the LOCAL method [Shenk *et al.* (1997) and Berzaghi *et al.* (2000)] use correlation between two objects as a measure of closeness. The CARNAC method, however, uses local averaging instead of local regression for prediction and also performs a data compression based on the Fourier transform (see Chapter 8.2) before the distances are computed.

## 11.4 Weight functions

In each local regression, the samples can be weighted according to how "far" they are from the

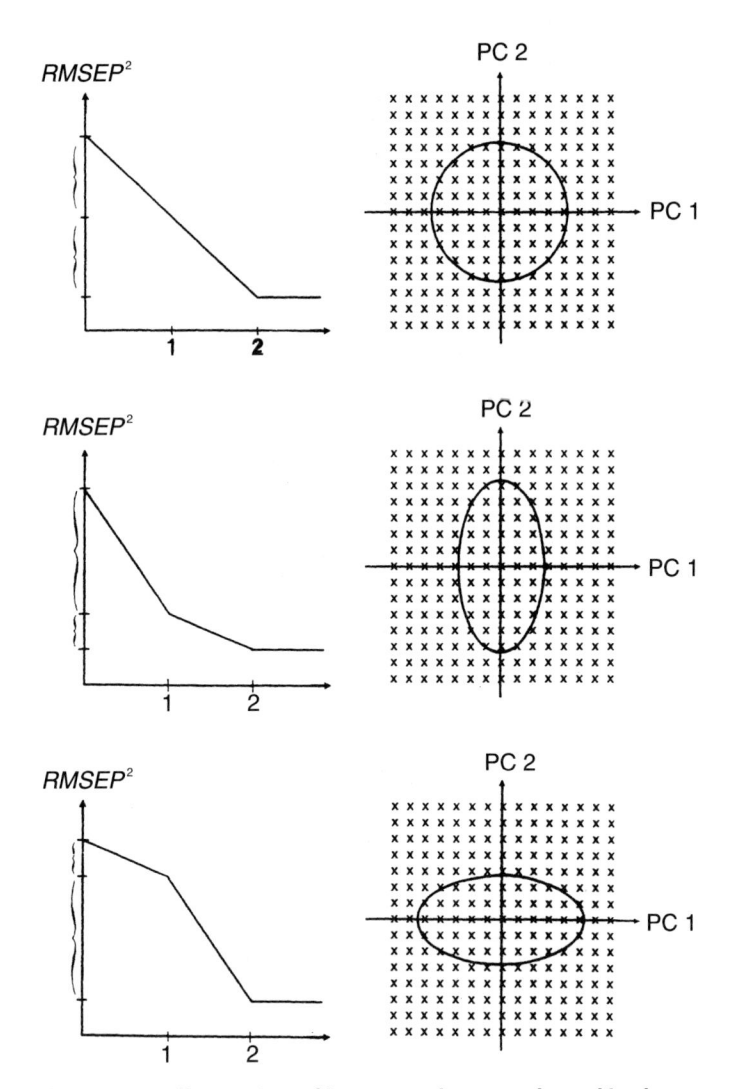

Figure 11.4. Illustration of how samples are selected in three different situations. The three illustrations on the left show situations where the relative importance of the two first components is very different. The figures on the right indicate how the modified Mahalanobis distance handles the three situations.

prediction sample. This can be done in different ways. Cleveland and Devlin (1988) proposed to use a cubic weight function defined by

11.4 Weight functions

$$W(d) = (1-d^3)^3, \quad 0 \le d \le 1 \quad (11.6)$$

where $d$ is the distance from the prediction sample to a calibration sample. The distance $d$ should be scaled in such a way that the distance to the calibration sample furthest away from the prediction sample within the local set is set equal to 1. This weight function is used in Næs et al. (1990) and in Næs and Isaksson (1992b). No clear conclusion exists about the relative merits of the two approaches, but it seems to matter less than the other decisions that have to be made [Næs and Isaksson (1992b)]. An illustration of the weight function in (11.6) is given in Figure 11.5.

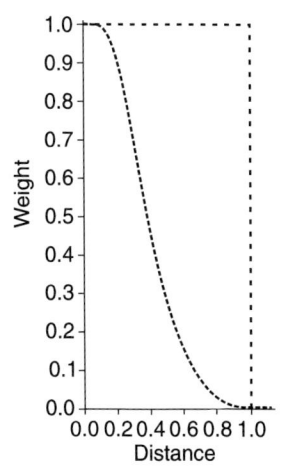

**Figure 11.5.** The cubic weight function described in equation (11.6).

## 11.5 Practical experience with LWR

The LWR method has been used in several applications and also compared to linear calibration methods such as PCR and PLS [Næs et al. (1990), Næs and Isaksson (1992b), Seculic et al. (1993), Wang et al. (1994) and Despagne et al. (2000)]. The general conclusion is that the method works well in comparison with alternative approaches. It has been demonstrated that improvements compared to linear methods of up to 30–40% are quite realistic even in relatively standard calibration situations. In addition, it has been observed that compared to PCR, the number of components needed for LWR to give good results is sometimes reduced substantially. This is important for interpretation purposes, but perhaps even more so for the stability and robustness of the calibration equations [Næs et al. (1990)]. The more components that are used in a calibration, the more sensitive is the equation to noise and drift.

In the last few years several papers have reported good results with the Shenk LOCAL method [Shenk, Berzaghi and Westerhaus (2000)]. This has given new interest to methods such as LWR and CARNAC that employ databases

# 12 Other methods used to solve non-linearity problems

## 12.1 Adjusting for non-linearities using polynomial functions of principal components

This chapter will describe other ways of using the principal components in non-linear modelling. This can, in principle, be done for any principal components, but here we will concentrate on the use of the first few components only. The main reason for this is that these components are the most stable and in many cases they contain most of the relevant information in the original $x$-data.

An illustration of a situation where the first two principal components contain the main information, but in a non-linear way, is given in Figure 12.1 (the same data are used to illustrate a related phenomenon in Figure 6.1). Figure 12.1(a) shows a design of a NIR experiment based on fish protein, water and starch. In Figure 12.1(b) the score plot of the first two components is presented (after MSC, Chapter 10). It is clear that the two plots resemble each other, but the fact that the level curves (contours) of the latter figure are not equidistant and straight indicates that the relationship between chemistry and NIR is non-linear. This is supported by the fact that an LWR model based on the two first components gives a root mean square error of prediction (*RMSEP*) which is more than 75% better than the *RMSEP* for the PCR based on the same two components and 33% better than the best possible PCR (with ten components).

**AA**
ANN: artificial neural network
GLM: generalised linear models
LS: least squares
LWR: locally weighted regression
ML: maximum likelihood
MLR: multiple linear regression
MSC: multiplicative scatter correction
NIR: near infrared
PCR: principal component regression
PCS: principal component shrinkage
PLS: partial least squares
PP: projection pursuit
*RMSEP*: root mean square error of prediction
WLS: weighted least squares

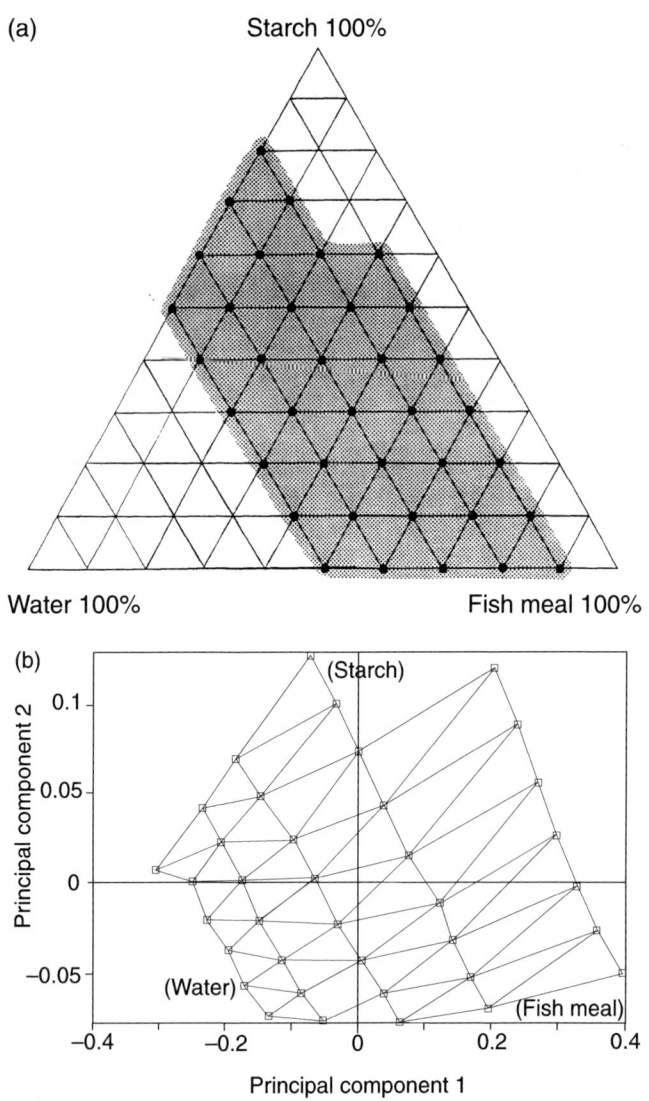

**Figure 12.1.** (a) The design of an experiment based on starch, fish protein and water. (b) The PCA scores plot of the NIR spectra of the samples in (a) (after MSC, see Chapter 10).

### 12.1.1 Polynomial modelling

A simple way of using the principal components is to combine them in a polynomial model, giving a

**12.1** Adjusting for non-linearities using polynomial functions of principal components

polynomial version of PCR. The same technique can be used for the original $x$-variables, but for collinear data a PCA is a reasonable pre-treatment before the actual non-linear modelling takes place.

If only two principal components are used, a second-degree polynomial regression model can be written as

$$y = q_0 + q_1 t_1 + q_2 t_2 + q_3 t_1^2 + q_4 t_2^2 + q_5 t_1 t_2 + f \quad (12.1)$$

Here the $q$s are the regression coefficients to be determined and $f$ is the random error term. A higher degree polynomial can also be constructed, but then more regression coefficients are needed. Note that the model is linear in the parameters, which means that it can easily be estimated by the use of standard linear regression using $t_1$, $t_2$, $t_1^2$, $t_2^2$ and $t_1 t_2$ as the five independent variables.

The method was tested in Isaksson and Næs (1988) and in Næs and Isaksson (1992a). The results in the first case were based on a second-degree polynomial in the first two components. The "non-significant" coefficients were identified by using projection pursuit (PP) regression [see Friedman and Stuetzle (1981) and Section 12.3.3]. The regression analysis was then performed only on the "significant" variables. The prediction ability obtained was comparable to the prediction ability of the best PCR with several more components in the model. In the second of the mentioned papers, similar results were observed.

If more than three or four components are used, the number of independent variables for this approach becomes large and the problem of over-fitting can sometimes become serious. Therefore, Oman et al. (1993) developed a technique based on added variable plots [see, for example, Weisberg (1985)] which can be used to identify the most important candidate variables. These plots are tailor made for detecting lack of fit of regression models and are based on plotting the

12.1 Adjusting for non-linearities using polynomial functions of principal components

dependent variable $y$ versus the candidate for addition, after first adjusting for the effects of the variables already in the model. Alternatively, regular variable selection procedures (Chapter 7) can be used.

### 12.1.2 An improvement based on Stein estimation

The polynomial PCR described above was improved in Oman *et al.* (1993) using so-called Stein estimation. The idea behind this method was to combine the non-linear information in the first few components with linear combinations of the rest of the components, which could possibly also be of some predictive relevance. The method shrinks the linear model based on all principal components towards a polynomial estimate based on the first few principal components. The method, which was named principal component shrinkage (PCS), is not computationally intensive and can easily be carried out using standard linear regression routines. The degree of shrinking is determined from the data.

The results using this method were found by Oman *et al.* (1993) to be slightly better than the results obtained by the standard non-linear PCR described above. More research is needed to evaluate the performance of the method.

## 12.2 Splitting of calibration data into linear subgroups

This method is based on splitting the calibration set into linear subgroups (subgroups within which it is reasonable to fit a linear regression) and performing linear calibration within each subgroup separately. The method was first published in Næs and Isaksson (1991), but is closely related to a similar method for segmentation in consumer studies published by Wedel and Steenkamp (1989) a couple of years earlier.

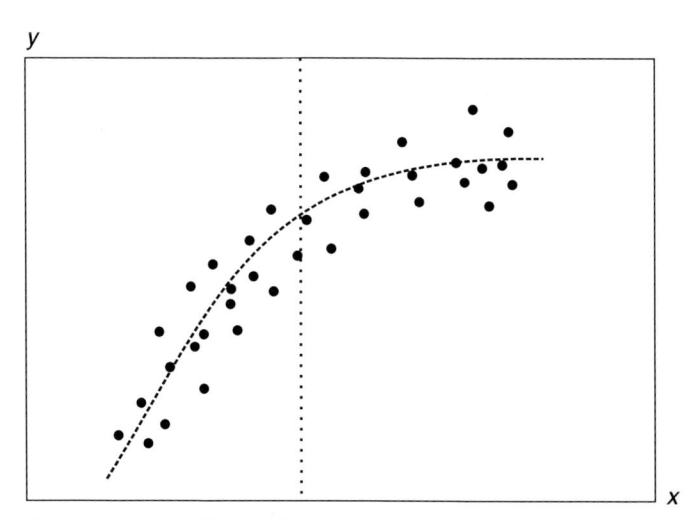

# Other methods used to solve non-linearity problems

Figure 12.2. An illustration of a situation where the data can be split up into two subgroups with reasonably good linearity within each of the subgroups. The split is indicated by the vertical dotted line.

The assumption behind the method is that the set of calibration samples can be split up in subgroups with an acceptable linearity within each of them. Each of the subgroups is then modelled using a linear calibration equation. New prediction samples must be allocated to one of the groups before prediction of $y$ can be carried out. This allocation can be made using a discriminant analysis method (see Chapter 18).

Note that this method is different from LWR in the sense that the present method develops "fixed" linear clusters to be used for the rest of the analysis. The present method may therefore be slightly easier to use in prediction since it is based on fixed linear models within each subgroup, but the calibration phase may be more time-consuming since identifying useful clusters may be quite difficult in practice.

## 12.2 Splitting of calibration data into linear subgroups

### 12.2.1 Cluster analysis used for segmentation

One possible way of splitting data into subgroups is by the use of cluster analysis (see Chapter 18). Usually, cluster analysis is based on the Euclidean or Mahalanobis distance [see, for example, Mardia *et al.* (1979)], but for the purpose of calibration Næs and Isaksson (1991) proposed to use residual distances obtained from linear modelling within each cluster. This gives clusters with as small residual sums of squares as possible within each of the subgroups. It was, however, observed that using the residual distance only, one could end up with samples far apart in space belonging to the same linear subgroup. Næs and Isaksson (1991) therefore proposed to use a distance which is a weighted average of a Mahalanobis distance in the (**x**,*y*) space and a squared residual from linear modelling within each subgroup. If we let $d_{ig}$ be the distance of object $i$ to group $g$, this combined distance can be written as

$$d_{ig}^2 = v d_{ig1}^2 + (1-v) d_{ig2}^2 \qquad (12.2)$$

where $d_{ig1}^2$ is a squared Mahalanobis distance in (**x**,*y*), $d_{ig2}^2$ is a squared residual and $v$ ($0 \leq v \leq 1$) is the parameter that balances the contribution from the two distances.

Note that this distance measure is only defined for objects relative to the subgroups. Hierarchical methods (see section 18.9.2), which require distances between all objects, are then not applicable. In Næs and Isaksson (1991) a so-called fuzzy partitioning method was applied (see Chapter 18). One of the nice features of fuzzy clustering is that it provides membership values (here called $u_{ig}$), which measure the degree of membership of each sample for each of the subgroups. These values can be useful for interpretation of the clusters.

Assume that the number of clusters is $G$, with clusters indexed by $g = 1, ..., G$, and as before the total number of calibration samples is $N$, with samples indexed by $i = 1, ..., N$. Then the membership matrix $\mathbf{U}$ is an $N \times G$ matrix of values between zero and one, showing the degree of membership of each of the $N$ objects to the $G$ clusters. The sum of the elements in each row of this matrix is assumed to be equal to one.

In fuzzy clustering as presented by Bezdec (1981), the number of groups $G$ is held fixed and the criterion

$$J = \sum_{i=1}^{N} \sum_{g=1}^{G} u_{ig}^{m} d_{ig}^{2} \qquad (12.3)$$

is minimised with respect to both $\mathbf{U}$ and $\mathbf{D} = \{d_{ig}\}$. The parameter $m$ in equation (12.3) can be chosen by the user. Small values of $m$ (close to 1) give "crisp" clusters with membership values close to either 0 or 1. Large values of $m$ give all membership values close to the average $1/G$. An $m$ equal to 2 seems to be a reasonable and useful compromise.

The algorithm for minimising the criterion $J$ is very simple. It alternates between two steps, one for optimising $\mathbf{U}$ for given $\mathbf{D}$ and one for optimising $\mathbf{D}$ for given $\mathbf{U}$. The algorithm usually starts with random $\mathbf{U}$ values. This algorithm has good convergence properties. For more information, see Chapter 18.

If the $\mathbf{X}$ data are highly collinear, the method can be applied to the principal components of the calibration set instead of the original $\mathbf{X}$.

### 12.2.2 Choosing G and v

After convergence of the fuzzy clustering algorithm, the different subgroups can be formed by allocating each object to the class for which it has the largest membership value. A separate linear regression is then performed within each group. The problem is then to determine the usefulness of the actual splitting, i.e. to find out whether the actual choice of $G$ and $v$ is a good one.

Inspection of **U** can be helpful in determining whether the actual splitting makes sense. For instance, a matrix **U** with many entries close to the average value, $1/G$, gives an indication that there are no clear subgroups in the data set. Windham (1987) proposed an index that can be computed from the **U** values, which can give a more direct and automatic tool for assessing the validity of the clusters. See also Kaufman and Rousseeuw (1990).

One can also use cross-validation or prediction testing for this purpose [see Næs and Isaksson (1991)]. The calibration procedure is run for a number of values of $G$ and $v$, the *RMSEP* is computed for each choice and the solution with the smallest *RMSEP* is selected. Preferably, the number of clusters $G$ should be as small as possible, so if two choices of $G$ give comparable results, the smaller should be selected.

### 12.2.3 Examples of local linear modelling

This method has, to our knowledge, so far only been tested in a relatively few applications. One of them is based on NIR analysis of protein percentage in samples composed of fish meal, water and starch [Næs and Isaksson (1991)]. Absorbance values at 19 wavelengths were scatter corrected by multiplicative scatter correction (MSC) and prediction samples were allocated to subgroups using the Mahalanobis distance (see Chapter 18). The number of calibration samples in this case was equal to 33 and the number of test samples was equal to 19.

Two groups ($G = 2$), two principal components ($A = 2$) and three different values of $v$ (0.0, 0.5 and 1.0) were tested. The prediction results (measured by *RMSEP*) for the different runs are shown in Table 12.1. As we see, the prediction results change quite dramatically according to which criterion is used. The results clearly show that $v = 0.5$ was better than $v = 1.0$ (Mahalanobis) and also slightly better than

Table 12.1. *RMSEP* values for different choices of *v*. In all cases G = 2 and A = 2.

|         | $v = 0.0$ | $v = 0.5$ | $v = 1.0$ |
|---------|-----------|-----------|-----------|
| Group 1 | 1.20      | 1.17      | 3.21      |
| Group 2 | 1.11      | 1.01      | 2.68      |

$v = 0.0$ (residual), showing that a compromise between the squared Mahalanobis distance and the squared residual was the preferred alternative. The best PCR result for the same data was equal to 0.91, but to obtain this, 15 components were needed. In other words, local linear modelling based on two principal components gave almost as good results as PCR with 15. As has been discussed before, using only a small number of components can be important for stability and robustness of the calibration equation.

Another example is from NIR analysis of moisture (transmittance, 850–1050 nm) in beef and pork samples [Næs and Isaksson (1991)]. There were 75 samples for calibration and 20 for testing. Three groups were used and $v$ was varied as above. The best results in this case were obtained for $v = 1.0$, showing that for this data set, the regular Mahalanobis distance gave the best result. The best *RMSEP* was (averaged over the three subgroups) equal to 0.89 using nine components. The details are given in Table 12.2. For PCR the best results were obtained for around 14 components and the *RMSEP* was equal to 1.19. Thus, local linear modelling gave an improvement in prediction ability and the number of components was reduced.

**12.2 Splitting of calibration data into linear subgroups**

**Table 12.2.** *RMSEP* **values for PCR and the splitting method. The average value corresponds to the total *RMSEP* for the calibration based on the splitting procedure. The number of principal components used is indicated in parentheses.**

| PCR | Group 1 | Group 2 | Group 3 | Average |
|---|---|---|---|---|
| 1.19 (14) | 1.10 | 0.53 | 0.86 | 0.89 (9) |

ANNs originate from Artificial Intelligence (AI) research and were the earliest models of how networks of interconnected neurons in the human brain produce intelligent behaviour. Current AI research uses much more sophisticated models but ANNs have found applications in many areas.

## 12.3 Neural nets

We will here describe the use of artificial neural networks (ANN) in multivariate calibration and also discuss their relationship to other methods such as partial least squares (PLS) regression and principal component regression (PCR).

For more theoretical discussions regarding approximation properties of ANN network models and statistical convergence properties of estimation algorithms, we refer to Barron and Barron (1988). For a general introduction to the ANN method, we refer to Pao (1989) and Hertz *et al.* (1991).

### 12.3.1 Feed-forward networks

The area of neural networks covers a wide range of different network methods, which are developed for and applied in very different applications. The "feed-forward" network structure is suitable for handling non-linear relationships between "input" variables $x$ and "output" variables $y$.

The input data ($x$) is frequently called the input layer and the output data ($y$) is referred to as the output layer. Between these two layers are the hidden variables, which are collected in one or more hidden layers. The elements or nodes in the hidden layers can be thought of as sets of intermediate variables analogous to the latent variables in bilinear regression (for example, PLS and PCR). An illustration of a feed-forward network with one output variable and one hidden layer is given in Figure 12.3. The information

# Other methods used to solve non-linearity problems

**Figure 12.3.** An illustration of the information flow in a feed forward neural network with one hidden layer.

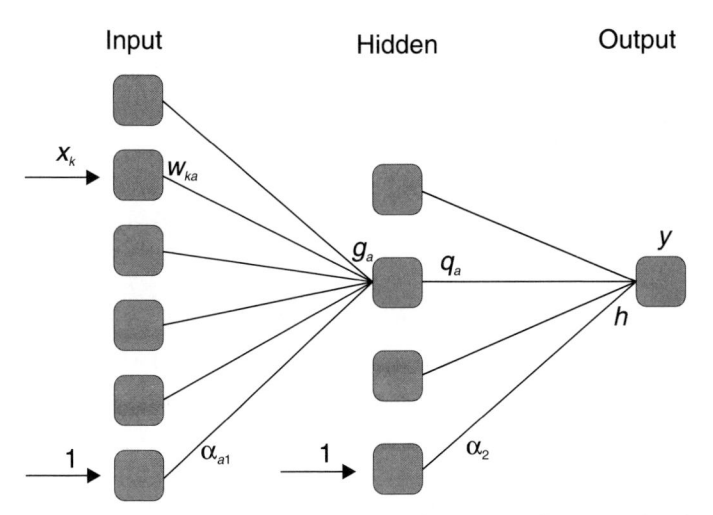

Each input is generally connected to all the nodes in the hidden layer. Most of these connections have been omitted in Figure 12.3 so that the connections for one input node can be seen more clearly.

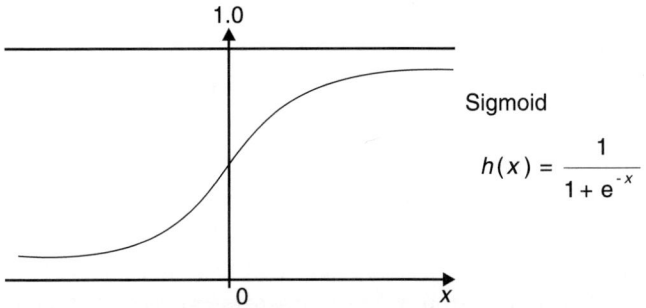

**Figure 12.4.** An illustration of the sigmoid function used within each node of the hidden and output layers.

$$h(x) = \frac{1}{1 + e^{-x}}$$

from all input variables goes to each of the nodes in the hidden layer and all hidden nodes are connected to the single variable in the output layer. The contributions from all nodes or elements are multiplied by constants and added before a non-linear transformation takes place within the node. The transformation is in practice often a sigmoid function, but can in principle be any function. The sigmoid signal processing in a node is illustrated in Figure 12.4.

## 12.3 Neural nets

The feed-forward neural network structure in Figure 12.3 corresponds to a regression equation of the form

$$y = h\left[\sum_{a=1}^{A} q_a g_a \left(\sum_{k=1}^{K} w_{ka} x_k + \alpha_{a1}\right) + \alpha_2\right] + f \quad (12.4)$$

where $y$ is the output variable, the $x_k$ are the input variables, $f$ is a random error term, $g_a(.)$ and $h(.)$ are specified functions and $q_a$, $w_{ka}$, $\alpha_{a1}$ and $\alpha_2$ are parameters to be estimated from the data. The constants $w_{ka}$ are the weights that each input element must be multiplied by before their contributions are added in node $a$ in the hidden layer. In this node, the sum over $k$ of all elements $w_{ka} x_k$ is used as input to the function $g_a$. Then, each function $g_a$ is multiplied by a constant $q_a$ before summation over $a$. Finally, the sum over $a$ is used as input for the function $h$. More than one hidden layer can also be used, resulting in a similar, but more complicated function. Note that for both the hidden and the output layer, there are constants, $\alpha_{a1}$ and $\alpha_2$, respectively, that are added to the contribution from the rest of the variables before the non-linear transformation takes place. These constants play the same role as the intercept term in a linear regression model. As can be seen from equation (12.4), an artificial feed-forward neural network is simply a non-linear model for the relationship between $y$ and all the $x$-variables. There are functions $g_a$ and $h$ that have to be selected and parameters $w_{ka}$ and $q_a$ that must be estimated from the data. The best choice for $g_a$ and $h$ can in practice be found by trial and error, but often sigmoid functions are used.

### 12.3.2 The back-propagation learning rule

In the terminology of artificial neural computing parameter estimation is called "learning". The learning rule that is most frequently used for feed-forward networks is the back-propagation technique. The objects in the calibration set are presented to the network one by one in random order, and the regression coefficients $w_{ka}$ and $q_a$ are updated each time in order to make the current prediction error as small as possible. This process continues until convergence of the regression coefficients. Typically, at least 10,000 runs or training cycles are necessary in order to reach a reasonable convergence. In some cases many more iterations are needed, but this depends on the complexity of the underlying function and the choice of input training parameters.

Back-propagation is based on the following strategy: subtract the output, $y$, of the current observation from the output of the network, compute the squared residual (or another function of it) and let the weights for that layer be updated in order to make the squared output error as small as possible. The error from the output layer is said to be "back-propagated" to the hidden layer. A similar procedure is used for updating weights in the next layer.

The differences between the new and the updated parameters are dependent on so-called learning constants or training parameters set by the user. Such learning constants are usually dynamic, which means that they decrease as the number of iterations increases. The constants can be optimised for each particular application [see Pao (1989) for details].

Neural networks are very flexible functions with many parameters to be determined. It is therefore extremely important to validate them properly (see Chapter 13). In some cases, it is advantageous to split the data into three subsets, one for fitting of the parameters, one for determining the architecture and one for validating the performance of the estimated network. For more information about this type of validation, we refer to Chapter 13.

Architecture
The number of hidden layers and nodes, and sometimes the number of connections.

**12.3 Neural nets**

Examples of the use of ANN in calibration of NIR instruments can be found in Næs *et al.* (1993) and Borggaard and Thodberg (1992).

### 12.3.3 Relations to other methods

In this section we will focus on the relation between feed-forward neural networks and other regression methods frequently discussed and used in the chemometric literature.

#### 12.3.3.1 *Multiple linear regression (MLR)*

Multiple linear regression (MLR) is based on the following model for the relationship between $y$ and $x$

$$y = b_0 + \sum_{k=1}^{K} b_k x_k + f \qquad (12.5)$$

where the $b_k$s are regression coefficients to be estimated. If both $h$ and $g_a$ in model (12.4) are replaced by identity functions, we end up with a model that is essentially the same as the linear model in (12.5). This is easily seen by multiplying the $w_{ka}$ and $q_a$ values and renaming the products.

Estimation of parameters in equation (12.5) is, however, much easier than estimation in the general neural network context. In Appendix A it is shown how the LS criterion leads to a simple closed form solution for the estimator.

#### 12.3.3.2 *Generalised linear models*

Generalised linear models [GLM, McCullagh and Nelder (1983)] are a class of methods which generalise the classical linear least squares theory. The methods extend the linear methodology by using a link function to relate the expectation of the $y$-variable to a linear function of $x$, and by allowing for error distributions other than the normal one. A neural network with only input and output layers and with a sigmoid (usually logistic) transfer function applied in the output layer can be regarded as such a generalised linear model. The only generalisation as compared to the linear model is that a sigmoid function is relating the expectation of $y$ to a linear predictor in the $x$-variables. In GLMs the parameters are usually estimated by using maximum likelihood (ML) estimation.

#### 12.3.3.3 Partial least squares regression and principal component regression

The regression equations for both PLS and PCR, based on centred $y$ and $\mathbf{x}$, can be written as

$$y = q_0 + \sum_{a=1}^{A} q_a \left( \sum_{k=1}^{K} w_{ka} x_k \right) + f = q_0 + \sum_{a=1}^{A} q_a t_a + f \quad (12.6)$$

where the $w_{ka}$s are functions of the loadings and loading weights [see Chapter 5 and Martens and Næs (1989)] and the $q_a$s correspond to the regression coefficients of $y$ on the latent variables $t_a = \Sigma(w_{ka} x_k)$. It can be seen that apart from the non-linear $g_a$ and $h$ in the feed-forward network (12.4), the two equations (12.4) and (12.6) are identical. This means that the model equation for PLS and PCR is a special case of an ANN equation with linear $g_a$ and $h$. Note that since PCR and PLS are based on centred $y$ and $\mathbf{x}$, there is no need for the constants $\alpha_{a1}$ and $\alpha_2$.

The learning procedure for PCR and PLS is, however, very different from regular back-propagation ANN. The idea behind the PCR and PLS methods is that the weights $w_{ka}$ are determined with a restriction on them in order to give as relevant and stable scores for prediction as possible. For PLS this is done by maximising the covariance between $y$ and linear functions of $x$, and for PCR it is done by maximising the variance of linear functions of $x$. In other words, estimation of $q_a$ and $w_{ka}$ for PCR and PLS is *not* done by fitting equation (12.6) to the $y$ data by LS.

Since back-propagation is based on estimation without restrictions, neural networks are more prone to overfitting than are PLS and PCR. As a compromise, it has been proposed to use principal components as input for the network instead of the original variables themselves [see, for example, Borggaard and Thodberg (1992)].

For PCR and PLS models, the scores $t_a$ for different $a$ are orthogonal. This is not the case for network models, in which the coefficients are determined without any restriction. The orthogonality properties of PLS and PCR help with interpretation of the data.

### 12.3.3.4 Projection pursuit (PP) regression

The PP regression method [Friedman and Stuetzle (1981)] is based on the equation

$$y = \sum_{a=1}^{A} s_a \left( \sum_{k=1}^{K} w_{ka} x_k \right) + f \qquad (12.7)$$

for the relation between $y$ and the $x$ variables. The difference between this and (12.6) is that the $q$s in (12.6) are regular scalars while the $s_a(.)$ in (12.7) are "smooth" functions [determined by local smoothing or averaging methods, see, for example, Friedman and Stuetzle (1981)] of the linear functions

$$t_a = \sum_{k=1}^{K} w_{ka} x_k \qquad (12.8)$$

Again, we see that the model (12.7) is similar to a network model with a linear function $h$ and the functions $g_a$ replaced by $s_a$. Notice also that since the $s_a$s have no underlying functional assumption, PP regression is even more flexible than the network in (12.4). Note also that since the $s_a$s are totally general except that they are smooth, we no longer need the constants $\alpha_{a1}$ and $\alpha_2$.

For the estimation of the weights $w_{ka}$ and functions $s_a$, PP regression uses a least squares strategy. First, constants $w_{k1}$ and a smooth function $s_1$ are determined. Then the effect of

$$s_1\left(\sum_{k=1}^{K} w_{k1} x_k\right) \qquad (12.9)$$

is subtracted before new constants $w_{k2}$ and a smooth function $s_2$ are computed. This continues until the residuals have no systematic effects left. For each factor, the $s_a$ and $w_{ka}$ are found by an iterative procedure. The functions $s_a$ are usually determined using moving averages or local linear fits to the data (see, for example, LWR as described in Chapter 11).

A drawback with PP regression is that it gives no closed form solution for the prediction equation, only smooth fits to samples in the calibration set. Therefore, prediction of $y$ values for new samples must be based on linear or other interpolations between the calibration points.

An application of PP regression in NIR spectroscopy can be found in Martens and Næs (1989).

*12.3.3.5 Non-linear PLS regression*

Frank (1990) published a non-linear PLS method, which is a kind of hybrid of PP regression and PLS. The technique is essentially based on the same model as PP regression, but the constants $w_{ka}$ are determined in a similar way as for PLS. Thus, this version of non-linear PLS is based on a similar, but more flexible model than the ANN model, but the estimation of coefficients is carried out under a restriction. The method is interesting since it contains an important aspect of flexibility in the modelling, but still uses restricted estimation. The method deserves more attention than it has been given so far. A related method is found in Wold (1992).

**12.3 Neural nets**

# 13 Validation

## 13.1 Root mean square error

After a calibration equation is computed, it is essential to determine its ability to predict unknown $y$-values. In particular, this is the case when choosing between alternative calibration methods and when deciding how many components, $A$, to use.

We will here concentrate on measures of prediction error based on squared differences between actual and predicted $y$-values. The quantity we will be focussing on is the root mean square error defined as

$$RMSE(\hat{y}) = \sqrt{MSE(\hat{y})} = \sqrt{E(\hat{y}-y)^2}$$

where $E(.)$ means statistical expectation (average) over the population of future samples. An advantage of using $RMSE(\hat{y})$ as compared to $MSE(\hat{y})$ is that it is given in the same units as the original measurements. Good calibration equations will have small $RMSE(\hat{y})$ values.

For some simple methods and applications, $RMSE(\hat{y})$ can be found using explicit formulae [see Oman (1985)], but usually it has to be estimated using one of the empirical validation methods to be described below. All these validation methods are based on testing the calibration on some test samples and using the test values to provide an estimate of $RMSE(\hat{y})$.

For validation, it is also useful to look at the plot of $y$ versus $\hat{y}$ for the samples used for testing. A good calibration will lead to observations falling close to a 45° straight line as shown in Figure 13.1. Such a plot has the advantage that it also can be used to identify regions with different levels of prediction accuracy. For instance, one will often see in such plots that $y$-values lower than the average are overestimated and

**AA**
ANOVA: analysis of variance
CV: cross-validation
DASCO: discriminant analysis with shrunk covariance matrices
MLR: mutliple linear regression
MSE: mean square error
MSEBT: bootstrap mean square error
MSEP: mean square error of prediction
PCR: principal component regression
PLS: partial least squares
RAP: relative ability of prediction
RMSE: root mean square error
RMSEBT: bootstrap root mean square error
RMSEC: root mean square error of calibration
RMSECV: root mean square error of cross-validation
RMSEP: root mean square error of prediction
SEP: standard error of prediction
SIMCA: soft independent modelling of class analogies

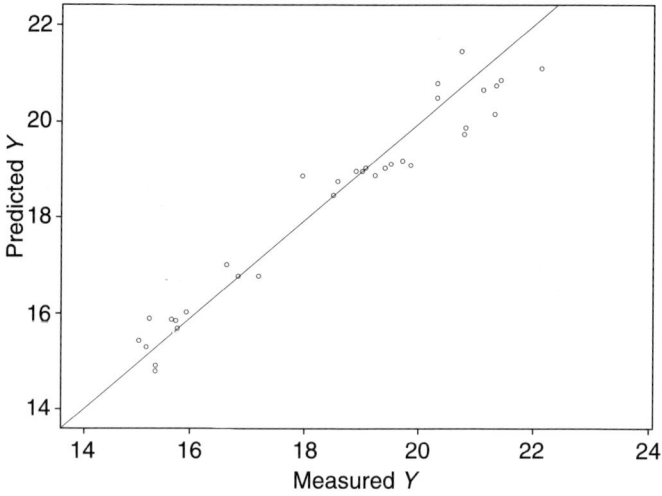

Figure 13.1. Plot of y vs ŷ for a NIR calibration tested on an independent test set. The y-value is protein content of meat. The NIR spectrum contains 100 wavelengths in the range 850–1050 nm (2 nm steps) and the calibration method used is PLS with seven components. There are 70 samples in the calibration set and 33 samples used for prediction testing. The best results are obtained for $A = 7$ with a *RMSEP* = 0.54 and a correlation equal to 0.97.

Some authors prefer to plot ŷ vs y, but in this book, we stick to the way presented in Figure 13.1. Both plots contain essentially the same information, the closer the points are to a 45° line, the better are the predictions.

that $y$-values higher than the average are underestimated (the least squares effect, see also Figure 3.1).

## 13.2 Validation based on the calibration set

An empirical estimate of prediction error used by some researchers is the root mean square error of calibration defined by

$$RMSEC = \sqrt{\sum_{i=1}^{N}(\hat{y}_i - y_i)^2 /(N - A - 1)} \quad (13.1)$$

where the ŷs are obtained by testing the calibration equation directly on the calibration data. The problem with this error estimate, however, is that it is essentially an estimate of the model error [$Var(f) = \sigma^2$, see Appendix A] and not of the prediction error. The estimation errors of the regression coefficients $\hat{\mathbf{b}}$ are not taken into account and *RMSEC* can be a highly over-optimistic estimate of the prediction ability. Especially for models with many *x*-variables or many PLS/PCR factors, the difference between *RMSEC* and true prediction error can be large.

In the following we will present some other and much more reliable techniques for estimating $RMSE(\hat{y})$.

## 13.3 Prediction testing

Prediction testing is based on splitting the data set into two, one for calibration and the other for validation/testing. The prediction testing estimate of $RMSE(\hat{y})$ is called root mean square error of prediction (*RMSEP*) and is defined by

$$RMSEP = \sqrt{\sum_{i=1}^{N_p}(\hat{y}_i - y_i)^2 / N_p} \qquad (13.2)$$

Here $\hat{y}$ and $y_i$ are the predicted and measured reference values for the test samples and $N_p$ is the number of samples in the test set.

An illustration of prediction testing is given in Figure 13.1 and its legend. The plot shows $y$ versus $\hat{y}$ for a number of test samples in an example of NIR analysis of protein in meat. The predictions are clearly good since all points lie very close to the straight 45° line. This is also reflected in a relatively small *RMSEP* value equal to 0.54 and a high correlation coefficient between $y$ and $\hat{y}$ equal to 0.97.

Prediction testing is the conceptually simplest validation method since *RMSEP* estimates the prediction ability of the **actual** predictor to be used, with all coefficient estimates already computed. A drawback is that several samples are set aside for testing only. These samples could instead have been used in the calibration set to give more precise regression coefficients with better prediction ability. Of course, the test samples can be put back into the calibration set after testing, but then the predictor is based on a larger set of samples and its properties are changed.

It is important to emphasise that a calibration equation will show different properties for different test sets. Therefore it is always important that the test samples cover the relevant range of samples as well as possible. A simple procedure that can be useful for selecting representative samples for testing is presented in Chapter 15. This procedure was originally developed for selecting calibration data, but it can also be used in this context. Likewise, the number of samples used for testing is also important. A rule of thumb, which should not be taken too rigidly, is to set aside about one third of the data for testing purposes and use the rest for calibration.

When prediction testing is used for determining the number of components to use in, for instance, PLS or PCR, a plot of the *RMSEP* versus the number of components may be very useful. Such a plot is illustrated in Figure 13.2. Typically, the *RMSEP* is large for small values of $A$, decreases as the number of components increases, before it increases as $A$ becomes too large. These effects correspond to the underfitting and overfitting effects as described in, for instance, Chapter 4. See also Figure 5.2 for an illustration of the same effects.

When using this type of *RMSEP* plot for determining the number of components to use in the future, one would normally search for the smallest *RMSEP* value. If, however, a smaller number of components gives approximately the same prediction error, one would normally select this smaller one. This solution contains fewer parameters and may be more stable in

# Validation

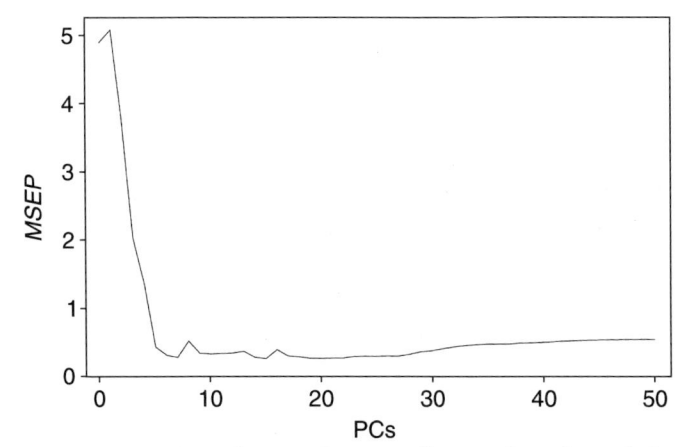

**Figure 13.2.** Plot of *MSEP* (=*RMSEP*$^2$) as a function of the number of components for the same NIR data as used in Figure 13.1. As can be seen, the *MSEP* is large for small values of *A*, then decreases to a minimum for six to seven components before it slowly increases.

the long run (see Section 13.7 for techniques that can be used to compare methods).

When the number of PLS/PCR factors, *A*, or the number of *x*-variables in stepwise MLR are determined by comparing *RMSEP* values, the test set becomes an important part of the calibration. In such cases, the *RMSEP* value of the best solution can sometimes give an over optimistic estimate of prediction error rate [see Nørgaard and Bro (1999)]. This is particularly true for very flexible methods such as neural nets. A safer, but also more data consuming strategy is to divide the test set into two sets, one for choice of model size and one for a more objective test of performance of the predictor. Another possibility is to combine prediction testing and cross-validation (see Section 13.4), use cross-validation for determining model architecture and prediction testing for testing the final performance of the method.

## 13.3 Prediction testing

In the legend for Figure 13.1, the correlation coefficient between $y$ and $\hat{y}$ was reported. The correlation coefficient is much used for validation, but it should be mentioned that it might in some cases be inappropriate. The reason for this is that correlation only measures degree of linear relationship between two measurements. A calibration equation that is clearly biased may still lead to a high correlation between measured and predicted $y$-values. In such cases, the *RMSEP* is to be preferred compared to the correlation coefficient. A plot of $y$ and $\hat{y}$ will reveal whether a calibration is biased or not. An illustration of a biased calibration equation is given in Figure 17.4.

## 13.4 Cross-validation

Cross-validation [Stone (1974)] is a validation technique based on the calibration data only. It is similar to prediction testing since it only tests predictors on data that are not used for calibration, but for cross-validation this is done by successively deleting samples from the calibration set itself. First, sample one in the calibration set is deleted. Then the calibration is performed on the rest of the samples before it is tested on the first sample by comparing $y$ with $\hat{y}$. The first sample is then put back into the calibration set, and the procedure is repeated by deleting sample two. The procedure continues until all samples have been deleted once. The estimate of $RMSE(\hat{y})$ based on this technique is called root mean square error of cross-validation (*RMSECV*) and is defined by

$$RMSECV = \sqrt{\sum_{i=1}^{N} (\hat{y}_{CV,i} - y_i)^2 / N} \qquad (13.3)$$

Here $\hat{y}_{CV,i}$ is the estimate for $y_i$ based on the calibration equation with sample $i$ deleted. As we see, this is very similar to *RMSEP* for prediction testing. The only difference is the way samples are set aside for testing.

An alternative to full cross-validation is to delete segments of samples. This runs faster on the computer since a much smaller number of calibrations have to be made. Since the statistical properties of a predictor depend on the number of samples used for calibration, this may, however, sometimes lead to a biased estimate of the $RMSE(\hat{y})$ for the predictor based on the full dataset. For full cross-validation where $N-1$ samples are used in each calibration, this is usually no problem since $N$ in most cases is large enough to ensure that $(N-1)/N \approx 1$.

Segmented cross-validation may be very useful if there are structures in the dataset. If, for instance, there are replicates of samples in the data, the $RMSECV$ computed by full cross-validation may give over-optimistic results. The reason is that none of the rows in the data matrix is unique, and therefore the same sample is represented in both the calibration and test set for each of the $N$ calibrations. In such cases, it is recommended that the replicates be forced to form a segment of samples to be deleted.

The $RMSECV$ must be interpreted in a slightly different way to $RMSEP$. Note in particular that it is not an error estimate of an actual predictor with already computed regression coefficients. Cross-validation provides instead an estimate of the average prediction error of calibration equations based on $N-1$ samples drawn from the actual population of samples. In technical terms, this means that for cross-validation both randomness in regression coefficients and the population of future samples is taken into account when computing the expectation in the definition of $RMSE(\hat{y})$. This implies that cross-validation is easiest to justify in situations with randomly selected calibration samples from a natural population. It can also be useful in other situations, but for designed experiments [for instance, fractional factorials, Box *et al.* (1978)], interpretation of *RMSECV* can sometimes be quite difficult.

The same plots as were used for prediction testing, Figure 13.1 and Figure 13.2, are useful also for cross-validation.

It should be mentioned that the problems discussed in Section 13.3 concerning the need for an extra test set to ensure a truly objective validation also hold for CV. If model architecture, like the number of components, is determined by CV, the absolutely safest is to test the predictor on a new and independent test set. Again, this is most important for the most flexible calibration methods, such as, for instance, neural nets.

The methods presented here and in the previous section are all focussing on predictions of *y* based on measurements of **x**. Similar methodology can be used to assess the dimensionality, i.e. the number of components of a principal component analysis of one single data table **X**. These analyses are often used for interpretation and in such cases an estimate of the number of significant, interpretable components may be very useful. Wold (1978) is an interesting reference for this methodology. We also refer to the discussion related to SIMCA and DASCO given in Chapter 18 of this book.

Regardless of how thoroughly a validation is done, it is always very important to monitor the calibration during regular operation (see Chapter 16).

## 13.5 Bootstrapping used for validation

Another empirical validation procedure which is useful is the bootstrap [see, for example, Efron (1982) and Efron and Gong (1983)]. The method is very versatile and can be used for a number of different purposes. One of them is estimating the prediction error of a calibration equation. It seems, however, to have been little used in practical applications of calibration procedures in chemistry.

# Validation

The bootstrap estimate of the root mean square error is obtained by first sampling $N$ observations from the dataset with replacement. The actual predictor is computed and a number of new samples are selected (again with replacement) from the dataset for testing. The average prediction ability of the test samples (mean square error of prediction, *MSEP*) is computed. The same procedure is repeated from the beginning a large number of times (for instance, more than 100) and the average *MSEP* is computed. This average can be called the bootstrap mean square error (*MSEBT*). If the square root is computed as was advocated above, we may call it *RMSEBT*. It is generally accepted that this is a good, but time-consuming method to assess predictor performance. It may be slightly biased, but its variance is usually better than for cross-validation [see, for example, Efron (1982)].

These ideas are not pursued further here, but more information can be found in Efron (1982), Efron and Gong (1983) and Efron and Tibshirani (1993). These publications contain the rationale for the method, theoretical details and also some of its properties.

## 13.6  *SEP, RMSEP, BIAS* and *RAP*

Another performance measure, much used in, for instance, NIR spectroscopy, is *SEP*, which is defined as the standard deviation of the predicted residuals

$$SEP = \sqrt{\sum_{i=1}^{N_P}(\hat{y}_i - y_i - BIAS)^2 /(N_p - 1)} \qquad (13.4)$$

Here

$$BIAS = \sum_{i=1}^{N_P}(\hat{y}_i - y_i)/N_P \qquad (13.5)$$

which can be interpreted as the average difference between $\hat{y}$ and $y$ in the prediction set. Note that since *SEP* is defined as a standard deviation, the number $N_p - 1$ is used in the denominator.

The relationship between *SEP* and *RMSEP* is simple:

$$RMSEP^2 \approx SEP^2 + BIAS^2 \qquad (13.6)$$

The only reason why exact equality is not obtained is that $N_p - 1$ is used in the denominator of *SEP* instead of $N_p$, which is used for *RMSEP*.

At this point it is appropriate to mention the distinction between accuracy and precision (see Figure 13.3). The concept *precision* refers to the difference between repeated measurements, while *accuracy* refers to the difference between the true and the estimated $y$ value. It is then clear that the *SEP* measures precision of a prediction, while *RMSEP* measures its accuracy. Our preference is therefore usually for the *RMSEP*. If *SEP* values are presented, one should always also report the *BIAS* (see also Chapter 17).

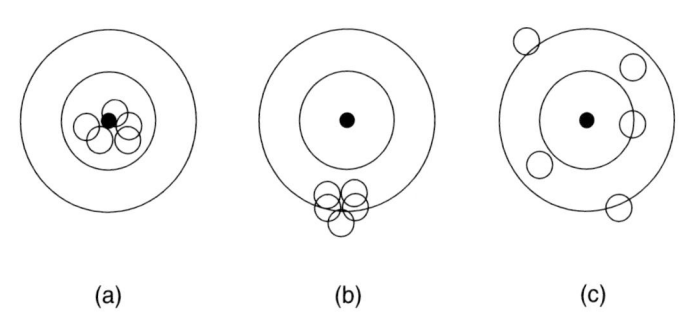

(a)     (b)     (c)

**Figure 13.3.** The precision and accuracy of a prediction illustrated. A predictor with high precision is one where the replicated measurements are close. Both situations depicted in (a) and (b) illustrate high precision. An accurate predictor is one with high precision and where the replicates are close to the true value. The situation in (a) is, therefore, one with high accuracy. The situation in (c) is one with neither high accuracy nor high precision.

Another criterion much used for validation is the correlation coefficient (see Appendix A) between measured and predicted $y$ value for the prediction samples, or for the cross-validated samples. This can be very useful and easy to interpret, but it may also have shortcomings as was discussed above. A variant of the regular correlation coefficient, the so-called relative ability of prediction (*RAP*) was defined by Hildrum *et al.* (1983). The *RAP* is an attempt to correct the standard correlation for the fact that test samples usually have *y*s with measurement noise in them (see also Chapter 13.9). We refer to Hildrum *et al.* (1983) for further discussion of *RAP*.

### 13.6.1 Example of the difference between *RMSEP*, *SEP* and *BIAS*

Let us assume that a calibration has taken place and that a linear calibration equation

$$\hat{y} = \hat{b}_0 + \sum_{k=1}^{K} \hat{b}_k x_k \qquad (13.7)$$

has been developed. Let us further assume that this equation is tested on seven randomly-selected samples with measured value of $y$. The $y$ and $\hat{y}$ measurements on these seven samples are as shown in Table 13.1.

The *RMSEP* is, in this case, equal to 0.41, the *BIAS* is 0.34 and the *SEP* is equal to 0.24. In other words, the *BIAS* is substantial compared to the *RMSEP*, and the *SEP* is about 40% smaller than the *RMSEP*. There is, of course, a certain chance that the calibration equation used has a real *BIAS* that should be corrected for, but with so few samples the chance of getting a substantial *BIAS* is always quite large. The *SEP* can therefore give an over-optimistic impression of the prediction ability.

**13.6** *SEP, RMSEP, BIAS and RAP*

Table 13.1. A data set used to illustrate the difference between *RMSEP*, *SEP* and *BIAS*.

| Obs. | $y$ | $\hat{y}$ | $\hat{y}-y$ |
|---|---|---|---|
| 1 | 5.2 | 5.7 | 0.5 |
| 2 | 7.3 | 7.4 | 0.1 |
| 3 | 2.3 | 2.2 | −0.1 |
| 4 | 4.8 | 5.3 | 0.5 |
| 5 | 8.4 | 8.9 | 0.5 |
| 6 | 6.1 | 6.5 | 0.4 |
| 7 | 3.7 | 4.2 | 0.5 |

## 13.7 Comparing estimates of prediction error

An important question when comparing calibration methods or selecting the number of components, $A$, is the following. How different do two error estimates have to be before we can be reasonably sure that the apparently better method really is better and the results wouldn't be reversed if we took another validation set? The methods to be discussed in this section can be used both for CV and prediction testing, but we will focus on the latter.

The situation is the following. Suppose two different prediction methods have been calibrated to predict $y$ from **x**. In the NIR context, $y$ is some lab measurement and **x**, usually multivariate, is spectral data. The prediction methods might be very similar, for example, two multiple regression equations, or very different, for example, a neural network with ten principal components as input and a simple regression equation based on the ratio of two derivative terms. The calibration may have been performed on the same or different calibration data. All that matters is that for a given **x**, each method will produce a prediction of $y$. Then suppose the two methods are compared by taking a single validation set of $N_p$ samples with known **x** and $y$ and by predicting $y$ from **x** using each of the methods. Since the true $y$ is known, this gives a set of $N_p$ prediction errors $(\hat{y} - y)$ for each method.

One way of summarising these results is to calculate the mean ($= BIAS$), the standard deviation ($= SEP$) or the root mean square error of prediction ($RMSEP$) of the $N_p$ errors. In the following we will first discuss a method for comparing $RMSEP$s (or $MSEP$s) before we discuss a related method for comparing $SEP$s.

The basic idea behind the method for comparing $RMSEP$s (or $MSEP$s) is that the squared errors for each sample and method constitute a two-way table with a structure similar to a two-way ANOVA table [see, for example, Searle (1971) and Lea et al. (1997)]. Thus, the model for the squared error is assumed to be

$$(\hat{y}_{ij} - y_{ij})^2 = \mu + \alpha_i + \beta_j + \alpha\beta_{ij} + e_{ij} \quad (13.8)$$

The index $j$ here refers to the method and $i$ to the sample. The symbol $\alpha_1$ is the effect of sample number $i$, $\beta_j$ is the effect of calibration method $j$, $\alpha\beta_{ij}$ is the interaction between sample and method and $e_{ij}$ is the random error. The sample effects $\alpha_i$ handle the fact that all methods are tested on the same samples. Without replicates in the model, the terms $e_{ij}$ and $\alpha\beta_{ij}$ are indistinguishable and the model becomes

$$(\hat{y}_{ij} - y_{ij})^2 = \mu + \alpha_i + \beta_j + e_{ij} \quad (13.9)$$

If test samples are randomly drawn from a population of samples, $\alpha_i$ and $\alpha\beta_{ij}$ should be treated as random ANOVA effects and therefore, $\alpha_i$ and $e_{ij}$ in model (13.9) can be considered random variables. In a mixed model without interactions, it is well known that hy-

**13.7 Comparing estimates of prediction error**

> The assumptions underlying the technique are usually easier to justify in prediction testing than in CV.

pothesis testing of fixed effects $\beta_j$s is identical to testing in a regular fixed effects model. Thus standard two-way fixed effects ANOVA without interactions can be used to test differences between the performance of calibration methods.

Note that a similar model would hold for any other suitable function of the residuals as well. In Indahl and Næs (1998), absolute values of prediction errors were used for this purpose. Note also that this technique can be used for any number of methods. If more than two methods are compared it may be necessary to use a multiple comparison method such as Tukey's test in order to find out where possible differences are. An example is given in Indahl and Næs (1998). The results in that paper show that differences between methods must be quite large before they are significantly different.

Significance testing in model (13.9) essentially requires that error terms are independent and that their distributions are reasonably normal. Indahl and Næs (1998) investigated the latter assumption for an application in image analysis. Residuals from a least squares fit of absolute values of prediction errors were checked and found to be fairly close to normal. The assumption of independent and normally distributed residual errors should preferably be checked in each particular application [see, for example, Weisberg (1985)].

The appropriate methodology to test differences between two *SEP* values can be found in Snedecor and Cochran (1967). The calculations go as follows. Find the correlation coefficient between the two sets of prediction errors and call it $r$. Then calculate

$$\kappa = 1 + \frac{2(1-r^2)t^2_{(N_p-2),\,0.025}}{N_p - 2} \qquad (13.10)$$

**13.7** Comparing estimates of prediction error

where $t_{(N_p-2),0.025}$ is the upper 2.5% percentile of a $t$-distribution with $N_p - 2$ degrees of freedom (the value $t = 2$ is a good approximation for most purposes). Then find

$$L = \sqrt{\left[\kappa + \sqrt{(\kappa^2 - 1)}\right]} \qquad (13.11)$$

Now

$$\frac{SEP_1}{SEP_2} \times \frac{1}{L} \quad \text{and} \quad \frac{SEP_1}{SEP_2} \times L \qquad (13.12)$$

give the lower and upper limits of a 95% confidence interval for the ratio of the true standard deviations. If the interval includes 1, the standard deviations (*SEPs*) are not significantly different at the 5% level.

Figure 13.4 presents $L$ as a function of $N_p$ for three values of $r$. The important message from this is that the intervals can be quite wide, when $N_p$ is modest and $r$ is not close to 1. For example, when $N_p = 30$ and $r^2 = 0.5$, then $L = 1.31$, i.e. the observed ratio of stan-

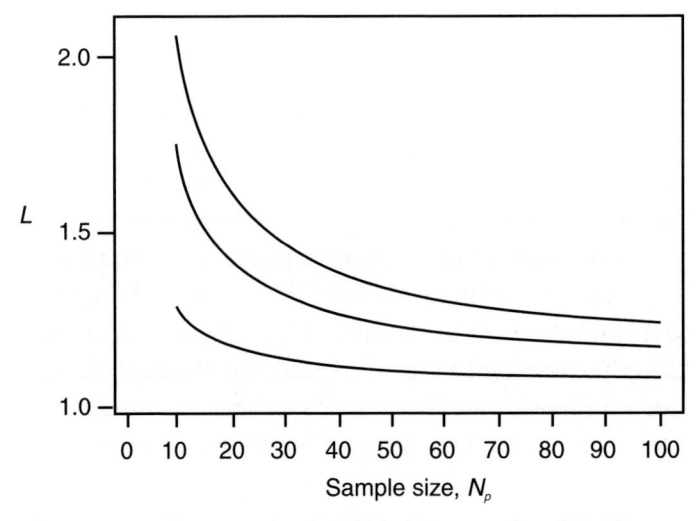

**Figure 13.4.** The quantity *L* defined in equation (13.11) as a function of $N_p$. The three curves, from top to bottom, are for $r^2 = 0$, 0.5 and 0.9, respectively.

## 13.7 Comparing estimates of prediction error

dard deviations needs to be greater than 1.3 (or less than 1/1.3) before we can claim a real difference with any sort of confidence.

## 13.8 The relation between *SEP* and confidence intervals

### 13.8.1 Confidence intervals

In standard statistical theory it is well known that a confidence interval is a stochastic (i.e. random) interval which with a certain probability contains the unknown parameter that we want to estimate. Such intervals are useful when determining not only the value of the unknown, but also to give an impression of the uncertainty in the estimation procedure.

In very simple cases with $N$ repeated measurements of a normally distributed variable, the confidence interval (with confidence coefficient 0.95) for the unknown mean can be written as

$$\bar{y} \pm 1.96 \times \sigma_{\bar{y}} \qquad (13.13)$$

where the standard deviation of the mean $\sigma_{\bar{y}}$ is assumed to be known in advance. If the standard deviation is unknown (which it usually is), $\sigma_{\bar{y}}$ must be replaced by its empirical analogue and the number 1.96 must be replaced by a value from a Student's t-distribution table [see, for example, Weisberg (1985)]. The t-table value is close to two for moderately large $N$.

This theoretical result is the basis for using the well-known "rule of thumb" for confidence intervals, namely estimated value ±2 times the standard deviation of the estimate. In many cases this is a reasonably good approximation to the exact confidence interval.

# Validation

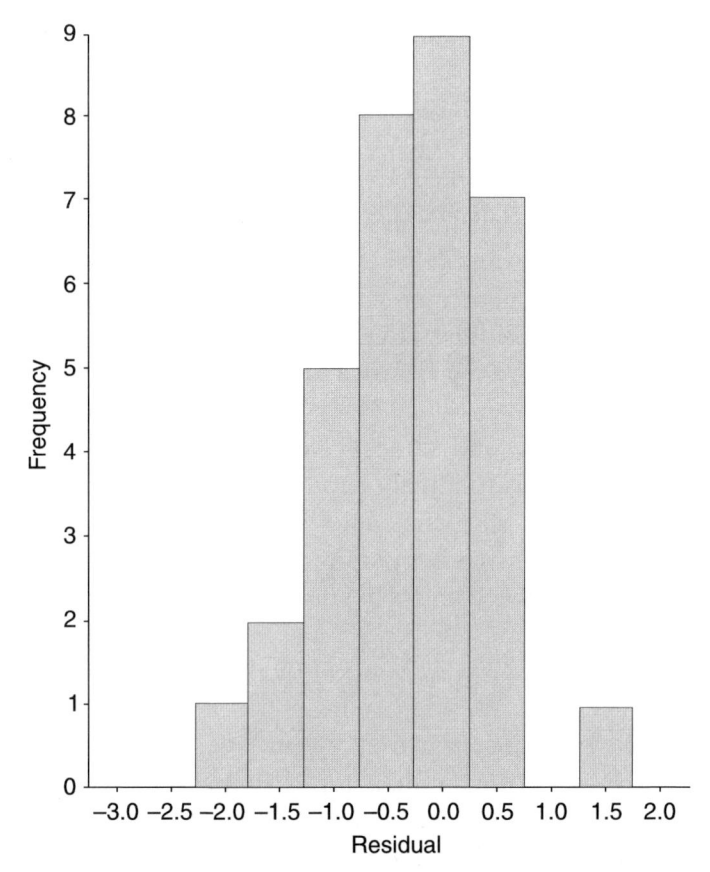

**Figure 13.5.** Histogram of prediction residuals from a NIR calibration.

### 13.8.2 Confidence intervals for predictions

Based on formula (13.13) it is clear that a natural suggestion for a calibration equation with small *BIAS* is to use

$$\hat{y} \pm 2 \times SEP \approx \hat{y} \pm 2 \times RMSEP \quad (13.14)$$

as an approximate 95% confidence interval for $y$.

Figure 13.5 shows a histogram of prediction residuals for a series of samples. The experiment is from moisture analysis in meat by NIR spectroscopy and the calibration method used is PCR. More details

**13.8 The relation between *SEP* and confidence intervals**

about the experiment can be found in Næs and Ellekjær (1992). As we see, the histogram is positioned with centre close to 0, which means that the *BIAS* is negligible in this case. As we can also see, the interval $(-2 \times SEP, 2 \times SEP) = (-1.5, 1.5)$ gives an adequate interval for about 95% of the data. This is therefore an example where the above confidence interval gives a good approximation to reality. If the approximation is unsatisfactory, bootstrap techniques can be used instead [Efron and Gong (1983)].

The interval in (13.14) represents an average length interval and is most relevant to use if the prediction error is relatively similar over the whole area of interest. Confidence intervals which depend on measurements of **x** are possible to develop, but this is quite complicated and is beyond the scope of the present book.

### 13.9 How accurate can you get?

It is not uncommon for NIR measurements to have better precision than the reference method. A commonly posed question is whether it is possible to have better accuracy of the NIR method compared to the actual reference measurement when used to determine the true reference value.

To answer this, let us assume that the method $x$ (for instance NIR) is perfectly reproducible, and would be perfectly accurate if only we knew the equation of the straight line that relates it to the true value. We calibrate against method $y$, which provides unbiased (the statistical expectation is equal to the true value), but rather imprecise measurements of the true value, with standard deviation $\sigma$. The calibration is performed by taking, for instance, four replicate measurements using method $y$ on each of two samples and a single measurement using method $x$ on the same two samples. The calibration graph is shown in Figure

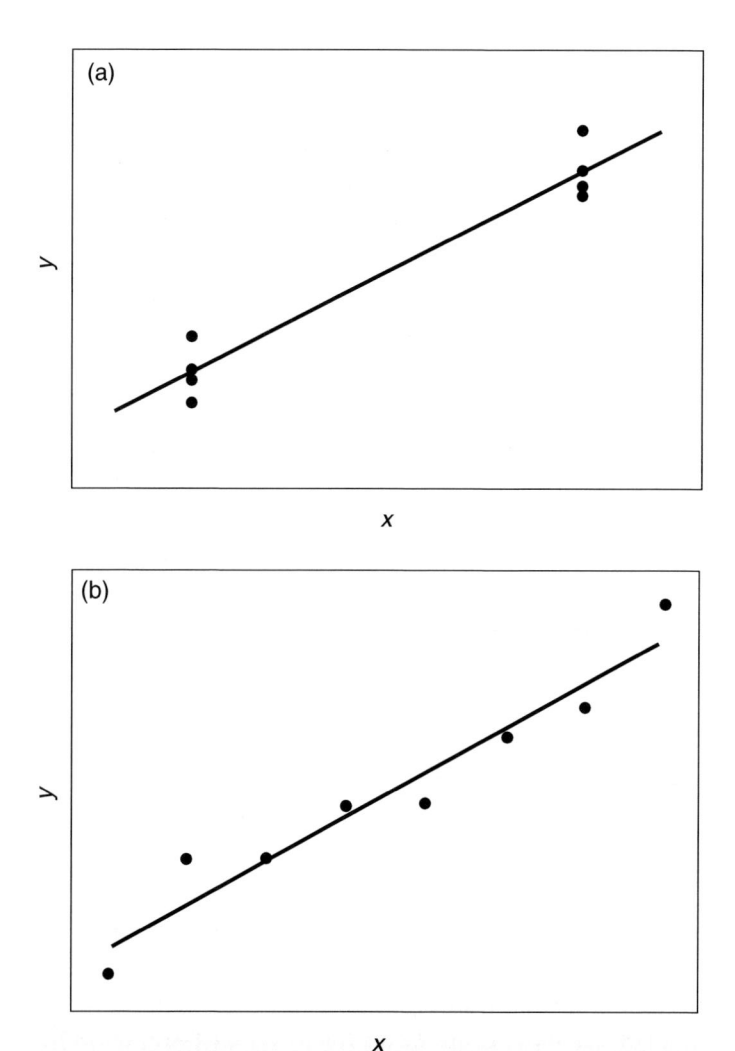

**Figure 13.6.** Conceptual illustration of the fact that one can obtain more accurate prediction by NIR than the reference method used or calibration. See text for an explanation.

13.6 (a), where the line passes through the two mean measurements.

We can now use $x$ and the calibration line to get a predicted $y$-value for a new sample. The only error in this measurement comes from the error in the calibration line caused by the errors in the $y$ measurements

**13.9** How accurate can you get?

used to estimate it. Some straightforward statistical calculations show that the error has a standard deviation of $\sigma/2$ or better everywhere between the two points that fixed it, i.e. whenever it is used for interpolation. Thus the method $x$ is now twice as good as the method it was calibrated against—although it costs us eight calibration measurements to achieve this.

It is clear that any desired accuracy could be achieved by increasing the number of measurements. Since $y$ is assumed unbiased, the two means through which the calibration line passes can be made to approach the true values as closely as we wish by using sufficient replication. What is less obvious is that we can achieve the same result by increasing the numbers of samples rather than by replicating measurements on the same samples as in Figure 13.6(b). By using statistical arguments it is possible to prove the same for this situation. The point is that the errors in $y$ average out; whether you use many replicates or many samples does not matter. In practice there are many reasons to favour lots of samples over lots of replicates (see also Chapter 15).

The over simplified set-up just described above can be complicated by adding some noise to method $x$ and making the regression multivariate, but the essence remains the same. If the reference method is unbiased, we can compensate for its lack of precision by using many calibration samples (or many replicates of a normal number of samples) and end up with a calibrated method whose accuracy is better than that of the reference. Note that these arguments assume that our underlying linear model is good. The accuracy of NIR to determine a reference value is not only limited by the noise in the reference, but also by the adequacy of the mathematical model. A calibration can never be better than the fit of the actual model used.

The problems start, however, when trying to assess the accuracy of the method or keep a check on its performance. All the regular ways of doing this involve comparisons of NIR results with reference results (see Sections 13.4 and 13.5 ). This is fine if the reference result can be regarded as close to the true value. If not, the reference error will also be part of the criterion used for validation. This was exactly the problem behind the development of *RAP* discussed in Section 13.6. If the noise in $y$ is assumed to be independent of $y$ itself, a noise-corrected version of *RMSEP* can easily be obtained by first taking the square of it, subtracting the noise variance and then taking the square root of the result. In mathematical terms, this can be written as

$$RMSEP(error\ corrected) = \sqrt{RMSEP^2 - \hat{\sigma}^2_{error}} \qquad (13.15)$$

where $\hat{\sigma}^2_{error}$ is the variance of the noise/error in $y$. This error estimate can be computed from replicated measurements of the reference value. We refer to Aastveit and Marum (1991) for an interesting application illustrating some of these points.

**13.9 How accurate can you get?**

# 14 Outlier detection

Mistakes and unexpected phenomena are unavoidable in the real world. This is also true in applications of chemometric techniques. There are always some observations that for some reason are different from the rest of the data set. In this chapter we will discuss some of the problems that such outliers can cause, how they can be detected and how they can be handled after detection. We refer to Cook and Weisberg (1982), Barnett and Lewis (1978) and Martens and Næs (1989) for more comprehensive discussions of outlier detection.

**AA**
LS: least squares
MLR: multiple linear regression
PCR: principal component regression
PLS: partial least squares

## 14.1 Why outliers?

In practice, there may be several reasons for an observation to be considered an outlier. One of these is when a sample, either a calibration or a prediction sample, belongs to another population than the "normal" samples. An example of this would be a barley sample that is included in a wheat data set. Such situations will result in a significantly different relationship among the $x$-variables than for the rest of the samples. For example, a barley sample would have a slightly different NIR spectrum to a wheat sample.

Another important case is when an instrument is not functioning properly and therefore gives an erroneous or misleading signal, affecting either one single $x$-variable or the whole collection. The effect of this will be the same as above, namely a different relationship among the $x$-variables.

A third important case is when there are errors in $y$ caused by reference method failure or a transcription error. Such an error will result in a sample which does not fit into the regression equation obtained from the rest of the data.

It follows from this that an outlier is not necessarily an erroneous observation, but merely an observation that is different from the rest and that can possibly have a strong influence on the results. Such observations can also represent new and valuable information for the researcher.

## 14.2 How can outliers be detected?

### 14.2.1 There are different kinds of outliers

#### 14.2.1.1 Univariate and multivariate outliers

Many of the outliers that are present in practical data sets can only be detected if several variables are considered simultaneously. Therefore, there are certain aspects of outlier detection that are easier to handle by multivariate techniques than univariate procedures. Figure 14.1 shows an illustration of this. A sample is presented in Figure 14.1(a) which is clearly lying far away from the regression line. If only $x$ is measured, as it is in prediction, and $x$ lies within the normal range, such outliers will pass unnoticed. Figure 14.1(b) illustrates a situation where two $x$-variables are measured and modelled. In this case, the new outlying sample is clearly detected as a sample that does not fit into the general variability of the $x$-variables, although both $x_1$ and $x_2$ fall within the normal range.

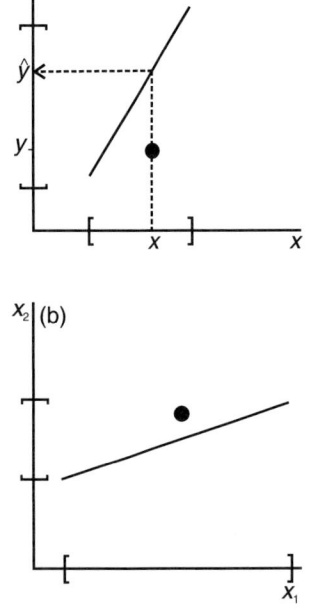

**Figure 14.1.** Multivariate methods can provide easier outlier detection than univariate regression. (a) Univariate calibration. (b) Multivariate modelling.

#### 14.2.1.2 Calibration and prediction outliers

It is important to distinguish between calibration outliers and outliers in the prediction phase. Calibration outliers are present in the calibration data set and therefore involved when estimating the prediction equation. If they pass unnoticed, they may possibly damage the whole equation with a consequence for all future samples. A prediction outlier on the other hand, is a sample measured after the calibration phase and will have no effect on the equation. The prediction of *y* for such samples will, however, usually be erroneous. Both situations depicted in Figure 14.1 relate to the prediction phase.

### 14.2.1.3 *x- and y-outliers*

It is also useful to distinguish between **x**- and *y*-outliers: the **x**-outliers are those **x**-vectors that in some way are abnormally positioned relative to the majority of **x**-data. This is relevant both for the calibration samples and for new prediction samples, since both types have a measured **x**-vector. The *y*-outliers are defined as those observations that have a different relationship between *y* and **x**. Since *y*-data are only available during calibration, *y*-outliers cannot be detected in the prediction phase.

In Figure 14.2(a), the outlying sample is an outlier in *y* close to the average in *x*. In this case the regression equation is only slightly influenced by this outlying sample. In Figure 14.2(b), however, we see that the outlying sample is abnormal both in *x* and in the relation between *x* and *y*. This sample has a very strong effect on the regression equation. We say that the sample is *influential*. In Figure 14.2(c), the sample is an outlier in *x* but is positioned close to the straight line fitted to the rest of the samples. This sample has little effect on the regression equation. It has been shown [see, for example, Cook and Weisberg (1982)] that in order to be influential, a sample has to be a certain distance away from the average of *x* and also be positioned away from the linear regression line fitted to the rest of the data. Technically speaking, a sample must have a contribution both from the leverage (distance in *x*-space, see, for example, Section 14.2.4) and from the regression residual in order to be influential.

**14.2** How can outliers be detected?

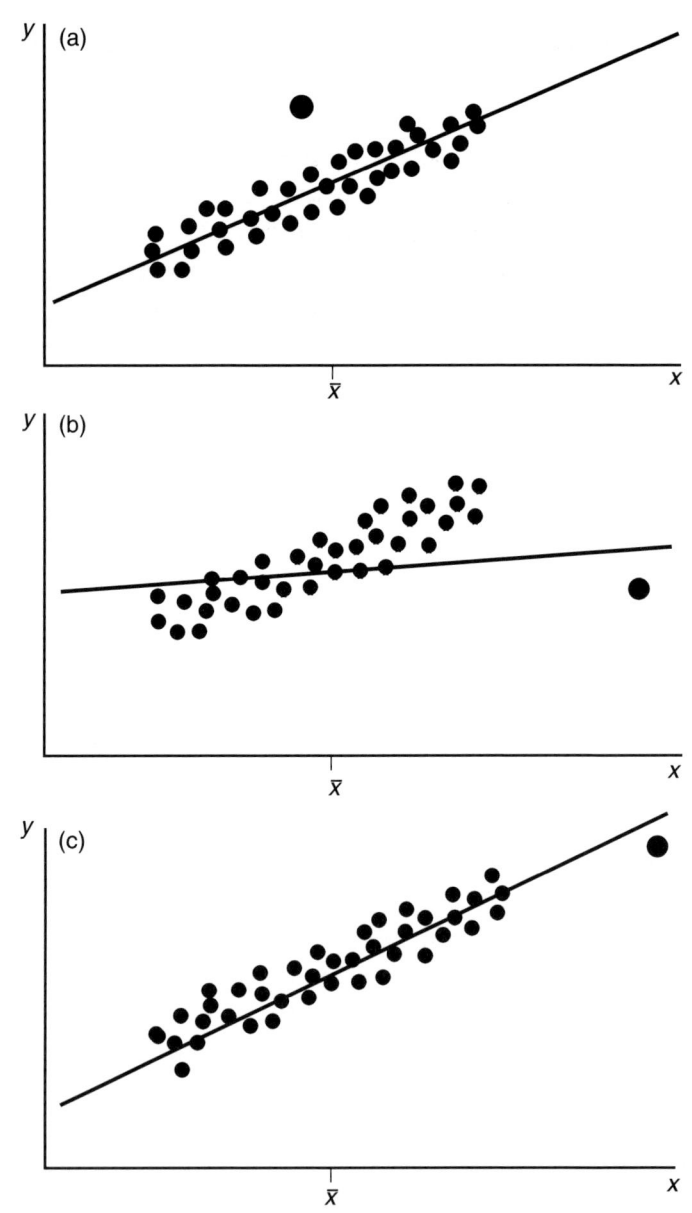

**Figure 14.2.** Outlier detection in linear models. The outlier in (a) is a y-outlier, the outlier in (c) is an x-outlier and the outlier in (b) is both an x- and a y-outlier. For calibration data, the observation in (b) is influential, the other two are not.

It is obviously most important to detect outliers that are influential.

### 14.2.2 Diagnostic tools that can be used to detect x-outliers

In one dimension, as in Figure 14.2, it is easy to define what it is meant by an observation being an **x**-outlier: one simply measures the distance from the sample to the average of all **x**-observations and uses this as a diagnostic tool. When it comes to more than one dimension, however, it becomes less clear what is meant by an observation (a spectrum for example) being an outlier. One can, of course, consider each variable separately and use the same approach as in the univariate case, but this approach does not take correlations into account and is not good enough for a proper handling of multivariate outliers. In the following we will concentrate on outlier detection tools related to bilinear regression methods such as PCR and PLS.

First of all, we will emphasise the importance of using plots of principal components scores and PLS scores for outlier detection. As was demonstrated in Section 6.4, these plots can be very useful for providing information about the data. If outlying samples are present, these will usually show up as points in the score plots lying outside the normal range of variability. Two- and three-dimensional score plots are most useful for this purpose. More formalised tools based on principal components and PLS components will be considered next.

#### 14.2.2.1 Leverage

The most important diagnostic tool to detect **x**-outliers is the so-called leverage, $h_i$ [see, for example, Cook and Weisberg (1982)]. As was indicated above, the leverage also plays a very central role in influence measures. Its name comes from its interpretation as the "potential" for a sample to be influential (see also Section 14.2.4).

The leverage is closely related to the so-called Mahalanobis distance [see Weisberg (1985) and Chapter 18.3.7] and is defined for sample $i$ as

$$h_i = \frac{1}{N} + \mathbf{x}_i^t(\mathbf{X}^t\mathbf{X})^{-1}\mathbf{x}_i \qquad (14.1)$$

Here $\mathbf{X}$ is the matrix of centred calibration $\mathbf{x}$-data and $\mathbf{x}_i$ is the centred $\mathbf{x}$-vector for sample $i$. Note that $\mathbf{X}^t\mathbf{X}$ is equal to $N-1$ times the empirical covariance matrix of the calibration $\mathbf{x}$s (see Appendix A). The contribution $1/N$ comes from the intercept in the linear model. The computed leverages can be plotted against observation number or against other diagnostic tools such as the $y$-residual (see below).

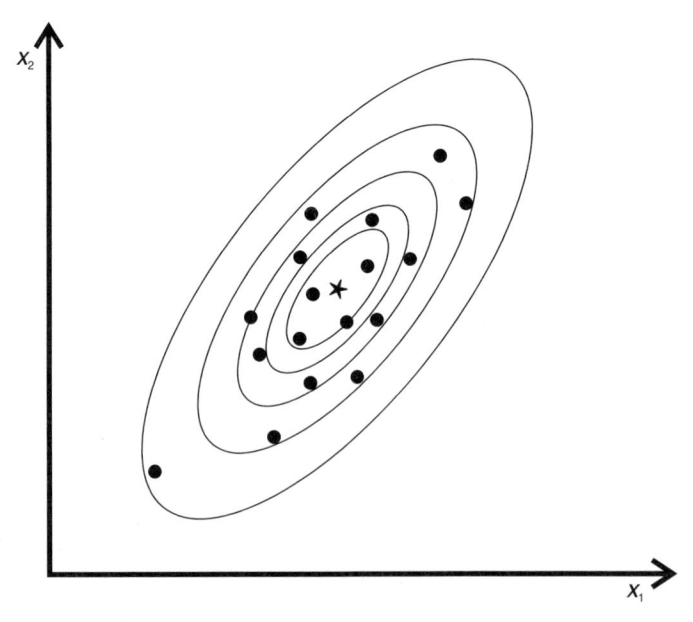

**Figure 14.3.** Points of equal leverage values are indicated by ellipses. The star in the middle is the average of all observations. The observations are given as filled circles.

Geometrically, points with equal leverage value are positioned on ellipsoids centred at $\bar{\mathbf{x}}$ as is illustrated in Figure 14.3. The equation $h_i$ = constant is the equation of an ellipse, whose shape is defined, via $\mathbf{X}'\mathbf{X}$, by the configuration of the calibration xs. A sample must be further away from the centre in a direction with large variability compared to directions with smaller variability in order to have the same leverage value. We can say that the leverage for an observation is its distance to the centre of all observations relative to the variability in its particular direction. When interpreted as a distance measure in $\mathbf{x}$ space, the leverage makes sense both for calibration and prediction samples.

The original leverage $h_i$ as defined in (14.1) is usually used in connection with MLR. When PLS or PCR are used for calibration, it is more common to compute $h_i$ only for the scores $\mathbf{t}_i$ that are used in the regression equation. Leverage computed for these methods is therefore a distance measure within the space determined by the $A$ components used, not within the full $\mathbf{x}$-space. Formally, the leverage of sample $i$ for methods with orthogonal scores (like PCR and PLS) is defined by

$$h_i = \frac{1}{N} + \sum_{a=1}^{A} \hat{t}_{ia}^2 / \hat{\lambda}_a \qquad (14.2)$$

Here $\hat{t}_{ia}$ is the score along component $a$ for sample $i$ and $\hat{\lambda}_a$ is the sum of squared score values for the scores of the calibration samples corresponding to component $a$. This sum of squares $\hat{\lambda}_a$ is (for PCR) equal to the $a$th eigenvalue of the matrix $\mathbf{X}'\mathbf{X}$.

Note that each component has its own contribution to the leverage. These individual contributions can also give useful indications of the influence of the sample $i$ on the estimation of the component directions themselves. For instance, if a sample has a very large leverage contribution for a particular compo-

**14.2 How can outliers be detected?**

nent, this is a strong indication that the actual sample itself is responsible for this one factor. If the sample were to be discarded from the analysis, this one factor would probably never show up.

For calibration samples, the leverage takes on values between $1/N$ and 1. The average leverage value for all samples in the calibration set is equal to $(1 + A)/N$. For prediction samples, $h_i$ can take on any value larger than $1/N$. No strict rules can be set up for when an observation should be called a "leverage outlier", but some rules of thumb are discussed in Velleman and Welch (1981). A possible warning may be given when the leverage is, for instance, two or three times larger than the average $(1 + A)/N$.

### 14.2.2.2 x-residuals

The x-residual vector is defined as the part of the **x**-vector that is not modelled by the components used in the regression (for instance, PLS or PCR components). In other words, the x-residual vector $\hat{\mathbf{e}}_i$ for sample $i$ is defined by

$$\hat{\mathbf{e}}_i^t = \mathbf{x}_i^t - \hat{\mathbf{x}}_i^t = \mathbf{x}_i^t - \hat{\mathbf{t}}_i^t \hat{\mathbf{P}}^t \qquad (14.3)$$

Here $\hat{\mathbf{t}}_i$ and $\hat{\mathbf{P}}$ are the vector of scores for sample $i$ and the matrix of x-loadings, respectively (see, for instance, model 5.1) and $\mathbf{x}_i$ is the centred **x**-vector for sample $i$. The residual vector can be visualised geometrically as the line segment between **x** and the plane in Figure 14.4. The plane corresponds to the two first principal components of the three x-variables and the residual defines the difference between **x** and the projection of **x** onto the two-dimensional space. The x-residual can, like the leverage, be computed both for calibration and prediction samples.

The x-residual vector $\hat{\mathbf{e}}_i$ can be used to detect observations that are not modelled by the same components/factors as the rest of the samples. Large residuals indicate that the sample contains other and

# Outlier detection

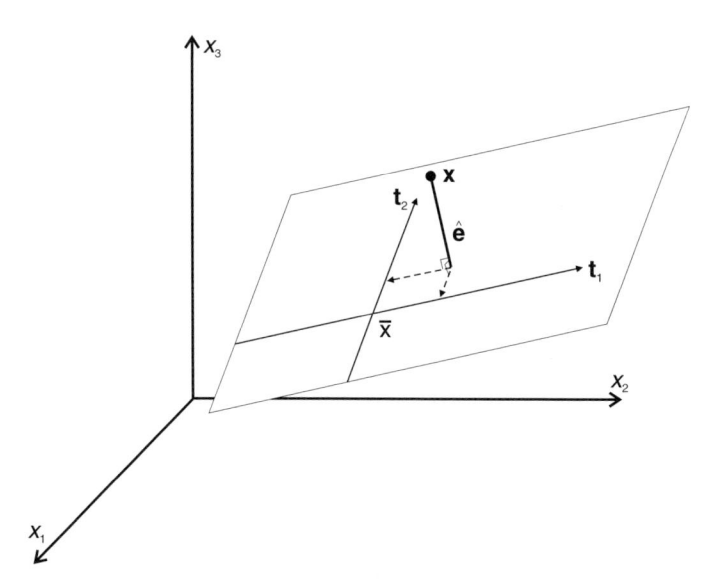

Figure 14.4. Geometrical illustration of the x-residual vector ê. This is the vector joining the observation x and the projection of x onto the plane generated by the model.

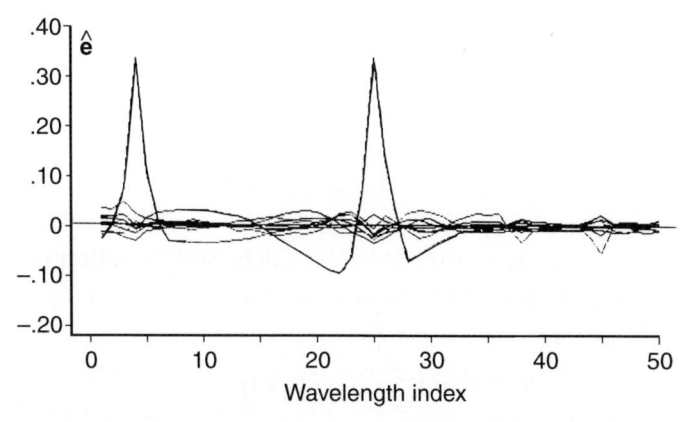

Figure 14.5. An x-residual plot. The plot is from a dataset of spectral measurements. Two of the spectra were artificially distorted by adding a spike to one of the wavelengths. Both of these spikes show up clearly in the residual plot. Reproduced with permission from H. Martens and T. Næs, *Multivariate Calibration* (1989). © John Wiley & Sons Limited.

## 14.2 How can outliers be detected?

unexpected constituents compared to the rest of the samples.

The residual vectors are best used in plots of residual values versus, for instance, wavelength number when the **x**-vector represents a spectrum. An example of such a plot is shown in Figure 14.5 [see also Martens and Næs (1989)]. In this case, two of the spectra look quite different from the rest. In some cases, the residual pattern can be used to give an indication of what has caused the change in spectral variability.

The sum of squared residual values for a sample gives a useful summary of the residual vector. Individual squared residuals of this kind can be compared to aggregated squared residuals (over samples) in F-tests to make more formal judgements about the degree of outlyingness [Martens and Næs (1989)].

*F-tests are described in Appendix A*

Since they contain information about different subspaces, it is useful to look both at residuals and leverages. This should always be done for the optimal number of components found in the validation, but a couple of other choices of $A$ (for instance, optimal plus and minus 1) should also be checked.

### 14.2.3 Diagnostic tools that can be used to detect y-outliers

*14.2.3.1 y-residuals*

The most common tool for detecting $y$-outliers is the $y$-residual. The $y$-residual for observation $i$ is defined as

$$\hat{f}_i = y_i - \hat{y}_i = y_i - \bar{y} - \hat{\mathbf{t}}_i' \hat{\mathbf{q}} \qquad (14.4)$$

i.e. as the difference between the actual $y$-observation and the $y$-value predicted from the model. The vectors $\hat{\mathbf{t}}_i$ and $\hat{\mathbf{q}}$ are, as before, the vector of score values and the vector of $y$-loadings, respectively. A graphical illustration of a $y$-residual is given in Figure 14.6. Here, the solid line corresponds to the regression model and

# Outlier detection

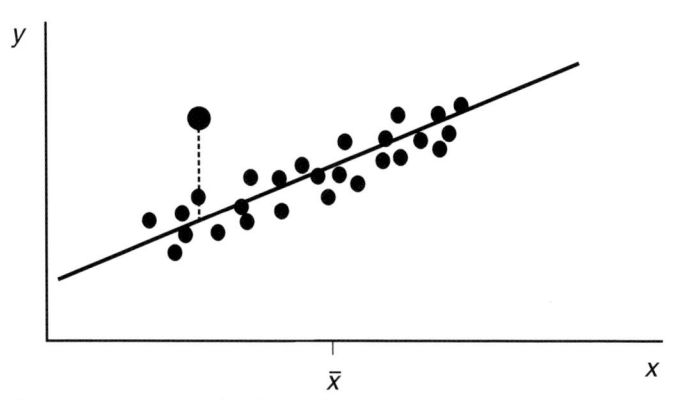

**Figure 14.6.** One point has a large vertical distance between itself and the regression line; it has a large residual.

the point corresponds to the actual observation. The dotted line segment between the point and the solid regression line is the $y$-residual.

The $y$-residuals are also best used in plots, either versus $\hat{y}$, versus one of the $x$-variables or versus observation number. There are no exact rules, but some rules of thumb can be set up for when a residual is to be considered larger than expected [see Martens and Næs (1989)]. A simple but useful warning limit is twice the residual standard deviation.

Large $y$-residuals can indicate errors in reference measurements or samples with a different relationship between **x** and $y$.

### 14.2.4 Influence measures

As was mentioned above, some outliers may be very harmful while others will have a very small effect on the regression results (Figure 14.2). There is thus a need for methods which can tell us whether an outlier is important or not. These tools are called influence measures [see, for example, Cook and Weisberg (1982) and Martens and Næs (1989)].

**14.2 How can outliers be detected?**

The most well known and probably the most used influence measure is Cook's influence measure [Cook and Weisberg (1982)]. It was defined directly as a measure of the influence of an observation on the regression coefficients, but it has been shown that it can also be written as

$$D_i = \frac{1}{K+1}\left(\frac{\hat{f}_i}{\hat{\sigma}\sqrt{1-h_i}}\right)^2 \left(\frac{h_i}{1-h_i}\right) \qquad (14.5)$$

As before, *K* is the number of variables in **x** and $\hat{\sigma}$ is the residual standard error.

This shows that an observation needs to have a contribution both from the *y* residual $\hat{f}_i$ and the leverage $h_i$ in order to be highly influential for the final regression equation. Note the correspondence with Figure 14.2, where only the outlying sample in (b) is influential; it has a contribution from both the residual and the leverage. The influence $D_i$ is usually plotted versus observation number. How to calibrate $D_i$ is discussed in Weisberg (1985).

Cook's influence measure was originally developed for MLR. Adaptations of the concept to regression analysis using data compression can be found in Næs (1989). For many applications it may be useful to plot the leverage (14.2) vs the *y* residual (14.4) for all the samples in the data set. This is called an *influence plot*. A sample with large values of both quantities is a sample with a large influence on the calibration equation. This tool can also be used for PLS and PCR.

### 14.2.5 Some other tools that can be useful for outlier detection

Leverage and residuals are the most important tools for detecting outliers. There are, however, some even simpler tools that can be useful and that should not be overlooked in practice. One of these simple tools which is useful in spectroscopy is graphical inspection of individual spectral values versus wavelength number. Such simple plots can reveal spikes and other unexpected phenomena that can indicate that something is wrong or at least that one should be cautious. Inspection of chemical values, checking whether they are within their natural range, checking whether replicates (for instance from different laboratories) are close etc., can also be extremely useful. Another technique that can be valuable is to check if the sum of the predicted constituents is 100%. This is of course only possible when all chemical constituents are calibrated for. All these simple checks may be performed using spreadsheet programs.

There may be cases when there are several outliers present in a calibration data set. Such outliers can, in some situations, mask each other in such a way that none of them shows up clearly using the regular diagnostic tools. Special methods have therefore been developed for handling such cases [Cook and Weisberg (1982)]. It may also be a sensible *ad hoc* strategy in such cases to delete the most dominant outliers first and then repeat the procedure until no significant outliers are detected, hoping that the less important will show up when the larger ones are deleted from the data set. A third possibility is to use so-called robust regression methods [Rousseeuw and Leroy (1987)], which are not influenced by outliers and therefore make it easier for some types of outliers to show up. Our advice is, however, that even when robust methods are used, outliers should be taken seriously because they can provide new and unexpected information.

## 14.3 What should be done with outliers?

When a calibration set outlier is detected, it is not easy to know exactly what to do about it, but some guidelines can be set up.

The first thing to do is always to go back to the lab and look for possible reasons for the sample to be an outlier. If something wrong is detected, one should of course correct the mistake if possible, or just leave the sample out of the calculations. If no clear reason for the outlier is detected, there are a couple of other possible ways to go. First of all, one should try to get an impression of the influence of the outlier on the final solution, using, for instance, the influence plot mentioned above. If the influence is negligible, it does not matter very much what is actually done with the outlier, whether it is kept as part of the calibration or left out of the calculations. The main problem occurs when the outlier is influential and no reason for the outlyingness is detected. *In such cases it is always difficult to know what to do*. If no other strong argument is available for keeping the sample, our advice is that the sample should be left out of the calculations (see also Appendix A). This will most often, but not always [Næs and Isaksson (1992a)] lead to better prediction ability of the predictor. If this solution is chosen one must be aware that there may be future samples similar to the outlier which will not fit to the calibration model and will simply not be possible to predict. The leverage and **x** residual described above for the prediction samples are important tools to help decide whether each new sample belongs to the population calibrated for or not. If a sample does not fit, accurate prediction is impossible and the sample should be analysed by another measurement technique.

# 15 Selection of samples for calibration

> **AA**
> LWR: locally weighted regression
> NIR: near infrared
> PCA: principal component analysis

To obtain good multivariate calibration equations one needs adequate statistical models, efficient estimation procedures and good calibration data. The earlier chapters have mainly focussed on model choice and how to perform estimation efficiently. The present chapter will discuss selection of useful data for calibration. We will again concentrate on linear calibration models

$$y = b_0 + b_1 x_1 + b_2 x_2 + \ldots + b_K x_K \qquad (15.1)$$

but similar principles apply when other types of model are used.

Usually, the more samples that are used for calibration, the better is the prediction ability of the equation obtained. It has, however, been demonstrated that it is not only the number of samples which is important [Isaksson and Næs (1990)], but also how the samples are selected. In the following we will discuss this in some more detail and also present some strategies for selecting "good" samples. It will also be emphasised that it is not only the prediction ability which is important, but also the ability to detect model problems and errors.

The main problem addressed by traditional experimental design methods [Box *et al.* (1978)] is to find settings of the explanatory variables **x** in order to obtain as good estimates of the regression coefficients as possible. In multivariate calibration situations, however, the **x**-values are usually some type of *spectral measurements*, which are functions of chemical and physical properties of the sample, and can then generally not be controlled. Therefore, standard textbook strategies can usually not be used directly. The ideas and principles lying behind the methods are, however, useful even in multivariate calibration as will be demonstrated below.

## 15.1 Some different practical situations

To make these ideas concrete, we will here describe two practical situations where classical experimental design methods cannot be used directly, but where ideas and principles behind them are still useful.

In the simplest practical situations it is possible to generate samples artificially in the lab. One will then have full control of the chemical composition **y** of the calibration samples. The **x**-data will usually be obtained as measurements using, for instance, a NIR spectrophotometer. This is a classical example of a situation where the **x**-data cannot be planned or designed, but where full control of **y** is possible. If design principles are to be used, they can therefore only be used for **y**, not for **x**.

A much more common practical situation is that the samples are natural products such as, for instance, meat. Such samples are very difficult, and often impossible, to make or construct in the lab. They must usually be taken as they are. This is, therefore, an example where neither the **x**-data, nor the **y**-data, can be designed. Using randomly selected samples for calibration is fine provided enough samples are used, but often the cost of the **y**-measurements is a limiting factor. Then there is a strong need for other strategies, where design principles are used to select smaller numbers of calibration samples that carry as much information as a larger random selection. One possible way of using principles from experimental design in such cases is to measure **x**-data for a large set of samples covering the region of interest as well as possible and then to select carefully, from the measured **x**-data, a smaller set of samples. These few selected samples are then sent to the lab for measurement of reference values **y**. The **x**-data are in many cases much easier to measure than the **y**-data (as they are in, for instance, NIR analysis), so such strategies are clearly realistic in practice. Design principles are in this case used for **x**, but only for selection of good candidates among a given set of measured samples.

## 15.2 General principles

In the two papers Næs and Isaksson (1989) and Isaksson and Næs (1990) the problem of selecting samples for both the above cases was considered in some detail and also illustrated by examples. The following three basic principles (or rules of thumb) were identified. As will be seen, the principles are fairly obvious and intuitive.

### 15.2.1 Basic principles

These are illustrated for two dimensions (though they can be generalised) in Figure 15.1 and explained in more detail below.

(1) All types of combinations of variables should be represented (i.e. there should be samples in all quadrants).
(2) The variation in all directions should be as large as possible, but limited to the region of interest.
(3) The calibration samples should be as evenly spread as possible over the whole region defined by 1 and 2.

For the first of the situations described above, the three principles apply to the design of the chemical

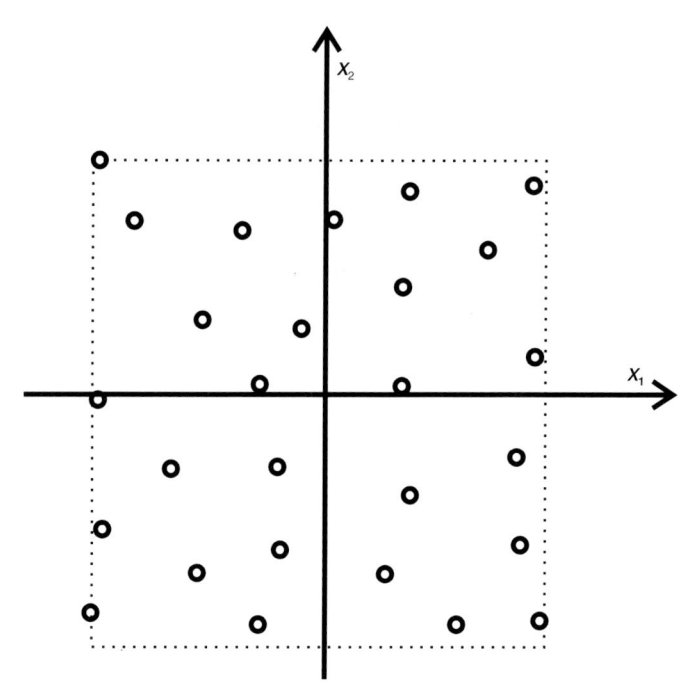

**Figure 15.1.** Principles for sample set selection. All combinations of variables are present, all directions are properly spanned and the samples are evenly distributed.

concentrations (**y**) and in the second case they apply to the most dominating directions (defined by, for example, PCA) of the **x**-data of the initial large data set.

The three principles above are illustrated in Figure 15.1 for a situation with only two variables, here denoted by $x_1$ and $x_2$. In the figure there are samples in all four quadrants. This means that all four types of combinations of the two variables are present (principle 1). Next, we see that samples are taken as far out as possible inside the region of interest indicated by the dotted line. This is principle 2. Finally, the samples are distributed evenly over the whole region, which corresponds to principle 3.

Principle 1 ensures that the calibration equation can be used for prediction of all *types* of samples

**15.2 General principles**

within the region of interest, not only for a few combinations of variables in a limited sub-region. The idea behind principle 2 is that a moderate range of samples is usually easier to fit by a linear model than a large range. In order to obtain a good model, one should then not use a larger region than necessary. Principle 3 ensures that possible non-linearities in the underlying relationship between **x** and **y** can be detected and in some cases also "smoothed out". The former is important for detecting problems with the model used and in some cases to indicate other possible solutions. The latter means that a linear model usually gives a better fit to non-linear data for such designs as compared to "end-point" designs, which are optimal when the linear model is true (see Figure 15.2). It has also been shown that a design with evenly distributed points over the region of interest is close to optimal, even when the linear model holds exactly [Zemroch (1986)].

Note that principle 3 is extremely important when local calibration methods such as LWR (see Chapter 11) are used.

In Næs and Isaksson (1989) an example was given of a situation where a design in **y** was possible. A number of different designs were generated and their effect on the quality of the calibration equation was investigated. It was demonstrated that the three principles above were important.

## 15.3 Selection of samples for calibration using x-data for a larger set of samples

Based on the three principles above, Næs (1987) suggested a procedure based on cluster analysis (see Chapter 18) for selecting calibration samples from a larger set of samples. The method was based on an idea by Zemroch (1986), but was modified for the

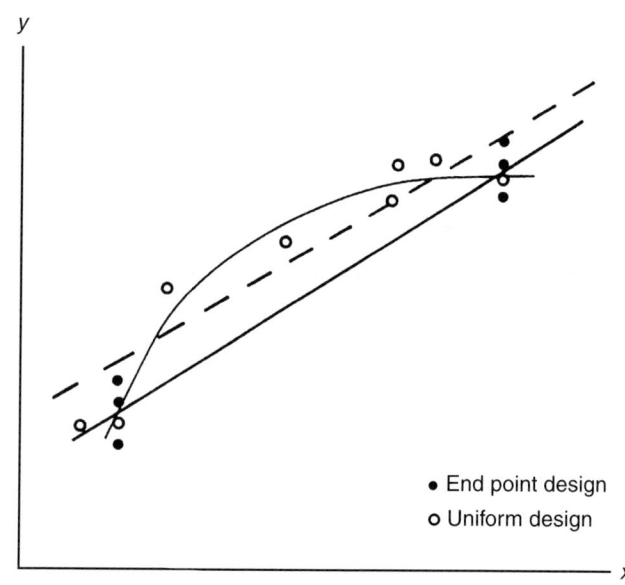

**Figure 15.2.** The principle of robustness for a design with evenly distributed points [see Box and Draper (1959) for similar ideas]. A linear equation is fitted to a curved relationship between $x$ and $y$. The end-point design gives a good prediction equation for the samples at the end (solid line), but very poor predictions in the middle. The evenly distributed design provides a reasonable linear model for the whole region even if the model is wrong. Note also that the even design can be used to check residuals over the whole region as opposed to the end-point design, which only has observations at the end. Reproduced with permission from T. Næs and T. Isaksson, *Appl. Spectrosc.* 43, 328 (1989) © Society for Applied Spectroscopy.

typical calibration situation with several collinear $x$-variables.

The following three points describe the method.

(a) Perform a PCA on all the **x**-data in the large data set (consisting of, for instance, 100 samples) and decide how many components are relevant for the constituent of interest. Typically, a small number of components is selected, for example, three to seven. Our experience is that the method is rather insensitive to this choice as long as it is not too small (less than three).

**15.3** Selection of samples for calibration using x-data for a larger set of samples

(b) Carry out a cluster analysis (Chapter 18) on the scores from the PCA. Stop the clustering when there are as many clusters as the number of samples one can afford to send for reference analysis. Several cluster analysis methods can be used. In Næs (1987), complete linkage was used.

(c) Select one sample from each cluster. These selected samples are the calibration samples, i.e. samples that are sent for analysis of their $y$ value. There are several ways to choose one sample from each cluster. One possibility is to use the sample within each cluster which is farthest away from the centre of the data is a whole, but other methods can also be envisioned.

The idea behind using PCA is to avoid the collinear dimensions of the data set. The idea behind the other two steps is that samples in the same cluster contain similar information and that duplication is usually not necessary. It is therefore better to select only one sample from each cluster, as opposed to using several samples from only a few of the clusters or from a limited region. In this way all three principles in Section 15.2 are reasonably well satisfied. All types of combinations of components are present (1), the samples span the whole space of interest (2) and the samples are evenly distributed (3).

In our experience, the method is easy to implement if standard statistical software is available and it seems to work well in practice. We refer to Næs (1987) and Isaksson and Næs (1990) for practical applications of the method. Good calibration results were obtained for sample sets of size down to about 20 samples, which is somewhat smaller than normally used. This shows that the *type* of samples may be more important than the *number* of samples used in the calibration.

Other and more manual procedures based on the same principles can also be envisioned. Using, for in-

**15.3 Selection of samples for calibration using x-data for a larger set of samples**

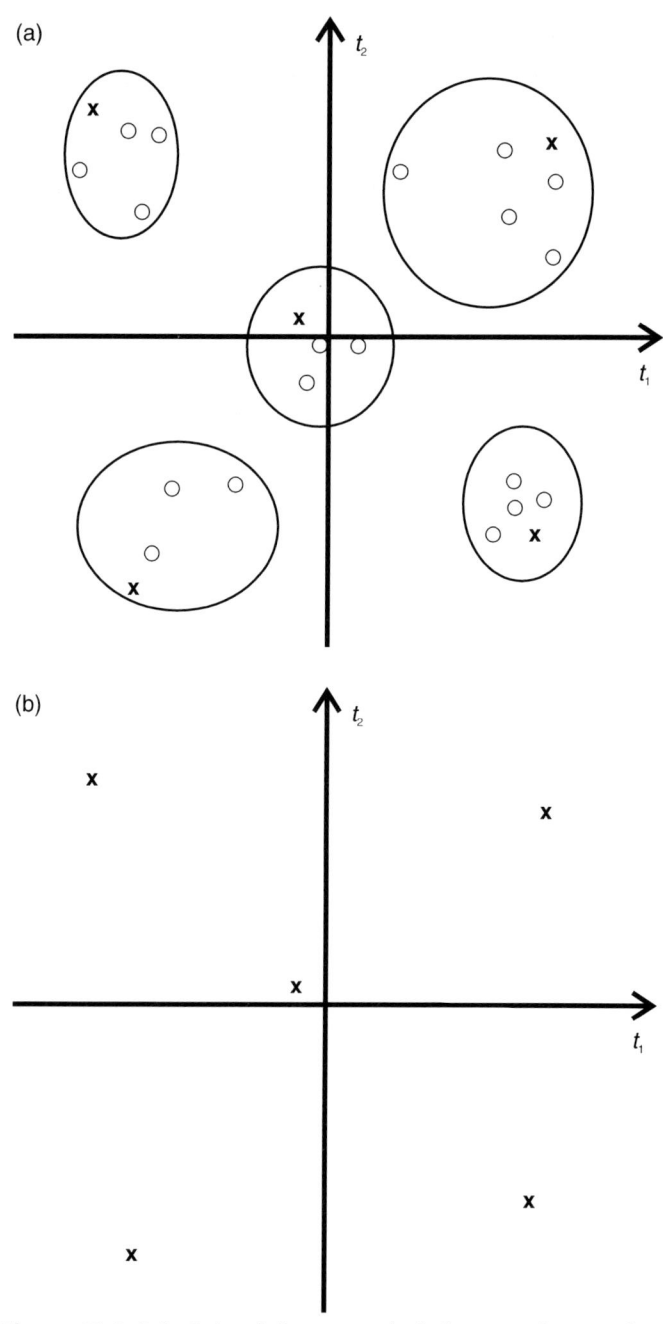

**Figure 15.3.** Principle of cluster analysis for sample set selection. In (a) is presented a set of points grouped into five clusters. In (b) one sample from each cluster is selected. These few samples are sent for chemical analysis.

### 15.3 Selection of samples for calibration using x-data for a larger set of samples

stance, visual inspection and selection of samples from the score plots of, for instance, the first three components can be an easy and very useful strategy.

We refer to Honigs *et al.* (1985) and Kennard and Stone (1969) for alternative strategies for sample set selection.

**15.3 Selection of samples for calibration using x-data for a larger set of samples**

# 16 Monitoring calibration equations

Given the current emphasis on accreditation, ISO 9000 etc., most analysts using NIR or other instruments in routine applications are likely to be monitoring the quality of the measurements in some way. Typically, this will involve checks on the instrument using standards or comparisons with wet chemistry determinations. When poor performance of a calibration equation is detected, one needs methods for modifying or improving the calibration performance. Such methods are presented in Chapter 17.

When monitoring is carried out it is important that a properly designed formal scheme is used. Informal ones can do more harm than good. In particular it is generally not a good idea to make regular small adjustments to calibrations as a result of measurements on just a few samples. The effect of this will usually be to increase the size of the error in the routine measurements, not to decrease it.

To see why, consider the idealised set-up where the instrument and the calibration are rock steady, so no adjustments at all should be made. Suppose that the measurement error, as assessed by the difference between a single instrument determination and a single lab determination, has a distribution with standard deviation $\sigma$. This is the standard deviation that is estimated by *SEP* calculated in the usual way on a validation set of samples. Now imagine a monitoring procedure in which one sample is put through the instrument and the lab each morning and the observed error for that sample is used to adjust all the measure-

---

NIR: near infrared
*SEP*: standard error of prediction
SPC: statistical process control

In situations where changes are expected to happen, when, for instance, a calibration is used for a new season or when samples are taken from another place, it is very important to check calibrations carefully.

Assume also that a good calibration equation has been made, so that the bias is negligible.

ments for that day in the obvious way by altering the bias in the calibration (see Chapter 17). The effect of this will be to increase the standard deviation of the errors considerably. This is because each reported measurement now has two error components: the error it would have had without adjustment, plus the (unnecessary) adjustment error.

What has just been described is a feedback control system that feeds back pure noise into the system it is trying to control. When the situation is idealised like this it is obvious that such a system is not a good idea. The problem is that any feedback control system, unless it is based on error-free measurements, is going to inject some extra noise into the system it is controlling. The best we can hope to do is to avoid feeding back any more noise than is necessary. This means not adjusting calibrations unless there is a good chance that more is gained in the bias adjustment than is lost by adjusting using measurements with some error in them.

A simple way to achieve this is to set up and run a control chart [Montgomery (1997)] using regular check samples, only making an adjustment to the calibration if the chart indicates that the system is clearly out of control. It is recommended that the chart is based on averages of several samples rather than on single measurements. This reduces error and improves the properties of the chart in detecting changes. It also enables gross outliers in the check samples to be spotted by comparison with the others in the subset. A typical recommendation is to use batches of three to five samples. This is based on the fact that the gain from averaging diminishes quickly for larger numbers. Using a mean of four samples halves the standard deviation, it takes 16 to halve it again and so on.

The samples do not all have to be taken at the same time. For example, one sensible scheme would

*This methodology is often called statistical process control (SPC).*

be to accumulate a weekly batch of five by selecting one sample at random from each day's throughput. The appropriate frequency of checks depends on the throughput and is a matter of balancing the cost of the checks against the possible losses from failure to detect a problem. If thousands of samples per week are being handled, daily checks might be appropriate. If there are only ten samples per week it does not make economic sense to check half of them.

The simplest type of chart, a Shewhart chart [Montgomery (1997)], is constructed by setting warning and action limits at $\pm 2\hat{\sigma}/\sqrt{N}$ and $\pm 3\hat{\sigma}/\sqrt{N}$ on either side of zero and plotting the mean errors for the batches of check samples sequentially as they arise, as in Figure 16.1. The $N$ in the two formulas above is the number of samples averaged on each occasion and $\hat{\sigma}$ is an estimate of the standard deviation of the error for each measurement. This needs to be constant if the chart is to work properly. It is usual to join the plotted points by lines, as in Figure 16.1, because this gives a

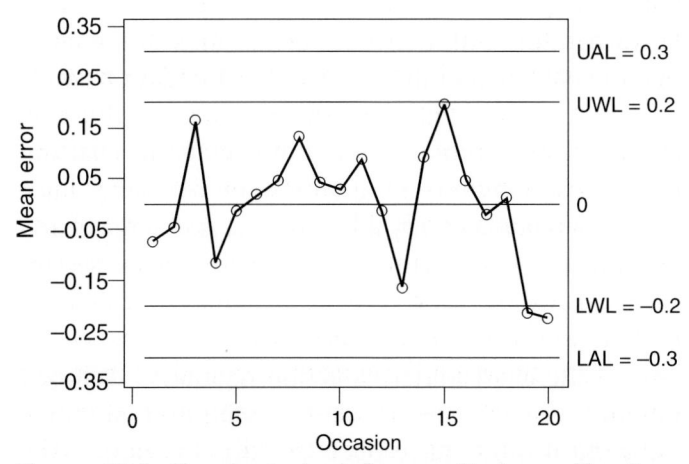

**Figure 16.1.** Shewhart chart for monitoring calibrations where $\hat{\sigma} = \sqrt{N} = 0.1$. UAL and LAL are upper and lower action lines, respectively. UWL and LWL are upper and lower warning lines, respectively.

better visual impression of any trends or patterns in the results.

The purpose of the warning and action lines is to show which average errors are within the expected range of variability and which are sufficiently large to indicate a possible problem. If the instrument (and the lab) are performing as they should, then only 1 in 20 points (5%) should plot outside the warning lines and roughly 1 in 1000 (0.1%) should plot outside the action lines. A usual rule is to take action if two points in a row plot between the same pair of warning and action lines or if any one point plots outside either action line. Thus the final two plotted points in Figure 16.1 would call for action. Some users add other rules for triggering action. Seven points in a row on the same side of the zero line is a popular one. Of course the more rules you add that cause an alarm, the higher the risk of false alarms. There are many introductory books on quality control that discuss these matters at length. Our feeling is that the details do not matter much, it is running the chart at all that matters.

If a detected change is sudden and substantial, it may indicate a problem with the instrument, with a new operator, with a grinder etc. A look at the chart may reveal whether there has been a trend, or may reveal that the problem appeared at the start of a new harvest, for example. The big advantage of charting the check samples, in addition to avoiding false alarms and unnecessary adjustments, is that the charts provide a visual record of the instrument's performance that show what is normal and may give clues to the timing and nature of problems.

If the chart indicates action regularly, then one possibility is that the standard deviation $\hat{\sigma}$ on which the control limits have been based is too optimistic. Calibration and validation exercises tend to be carried out under tighter control than routine operation. Validation samples are often obtained by setting aside part of the calibration set. It should not then be a surprise if the validation samples are more like those in the calibration set than are those encountered in routine operation. The sign that the limits are set unrealistically tightly is a chart that shows what looks like random fluctuations (as opposed to trends or a tendency to stay on one side of zero) but regularly crosses the control lines. The right course of action then is to reset the lines, using a standard deviation calculated from the sequence of check samples, and to revise one's opinions on the capability of the system. A wrong course of action may be to make regular adjustments to the calibration in the mistaken belief that this is improving the situation.

# 17 Standardisation of instruments

## 17.1 General aspects

The fundamental problem of standardisation is to make an instrument response conform to a "standard" instrument response. Of course, most of this standardisation work is done by hardware technology and taken care of by instrument manufacturers, but experience has shown that even instruments of exactly the same type may be different enough that an extra mathematical standardisation is needed. In particular, this is true for many NIR instruments, for which a number of useful procedures for mathematical standardisation have been developed. This chapter will describe some of these procedures.

### 17.1.1 Reasons for standardisation

In practice, there may be several reasons for instruments to need a standardisation. Typical examples are given below.

### 17.1.2 Different instruments or measurement modes

A common situation in routine analyses is that one wishes to use the same calibration on several instruments. In larger companies there may, for instance, be a need for a network of several instruments or one may be interested in transferring a calibration from an off-line laboratory instrument to an on-line process analyser. Such a transfer may also include a change of the measurement mode, from, for example, sample cup measurements to fibre optic measurements or from reflectance to transmittance measurements. It can also involve instruments of different types (monochromator, FT-NIR, filter, diode

**A**
DS: direct standardisation
MLR: multiple linear regression
MSC: multiplicative scatter correction
NIR: near infrared
PDS: piecewise direct standardisation
PLS: partial least squares
*RMSEP*: root mean square error of prediction

Many of the methods reviewed in this chapter were developed specifically for use with NIR instruments, and are described in this context. They may, of course, also find application elsewhere.

array, light emitting diodes, AOTF etc.) or other less fundamental instrumental differences.

### 17.1.3 Instrument drift

All measurement equipment will change with time, due to wear, drift etc. and needs service, cleaning and adjustments that may alter the measurement results.

### 17.1.4 Change in measurement and sample conditions

Sometimes changes in physical measurement conditions, such as humidity or temperature, may cause changes in spectral measurements. Minor changes in the composition of a sample, due to, for instance, change of grinder or homogeniser, can also give rise to a need for standardisation.

We will concentrate here on some of the simpler and probably the most used methods for instrument standardisation. Other possibly useful approaches that will not be discussed here are standardisation based on the wavelet transform [Walczac *et al.* (1997)], the patented algorithm by Shenk *et al.* (1985) for NIR calibrations, transfer by the use of orthogonal signal correction [OSC, Wold *et al.* (1998)] and the use of neural nets as described in Despagne *et al.* (1998). If none of these techniques work, the final resort is to do a full recalibration of the instrument using new samples and new spectral readings.

Interesting and useful overviews are given in de Noord (1994) and Nørgaard (1995). Other useful references are Bouveresse *et al.* (1994), Bouveresse and Massart (1996), Forina *et al.* (1995) and Adhihetty *et al.* (1991).

Before we describe some of the most useful methods we must define some concepts: we define instrument $\underline{A}$ to be the instrument where the calibration was originally made (master) and instrument $\underline{B}$ to be the instrument where the calibration is to be transferred (slave). A spectrum measured on instrument B̲ is denoted by $x_B$, a spectrum measured on instrument A̲ is denoted by $x_A$, and a spectrum on B̲ after some transformation designed to make it resemble a spectrum from A̲ is denoted by $x_{B(A)}$.

## 17.2 Different methods for spectral transfer

### 17.2.1 Spectral transfer methods with the use of standard samples

This group of methods is probably the most used in practice. It works by measuring a number of standard samples on both instruments A̲ and B̲. The standard samples should be as stable as possible during measurement and should also be similar to the samples of the actual application. This ideal is often difficult to attain in practice for biological materials.

The original data for all standard samples from the two instruments can be collected in matrices which here will be denoted by $X_A$ and $X_B$, respectively. As usual each row of the matrix is the spectrum of a sample. These matrices represent the information used to find the relationship or transform between the two instruments. If a linear transfer method is used, the transform corresponds to multiplying each spectrum $x_B$ by a matrix $B$ to obtain $x_{B(A)}$.

#### 17.2.1.1 Direct standardisation (DS)

The direct standardisation techniques can be split into the class of multivariate and the class of univariate techniques. The multivariate approach assumes the model

$$X_A = X_B B + E \qquad (17.1)$$

Here $B$ is a matrix of unknown parameters and $E$ is the residual matrix. If desired, an intercept can be incorporated. This is particularly important if a baseline difference is expected. The unknown **B** matrix is estimated by using, for example, PLS2 regression as described briefly in Section 5.4. After estimation of the matrix **B**, a new sample can be transformed by

$$\mathbf{x}'_{B(A)} = \mathbf{x}'_B \hat{\mathbf{B}} \qquad (17.2)$$

Note that both the variability between instruments and the variability between the standard samples is modelled in equation (17.1). Therefore, one will always have a situation where a lot of the variability in $\mathbf{X}_A$ can be accounted for by $\mathbf{X}_B$. The important point is therefore whether the estimated residual matrix **E** in formula (17.1) is smaller than the difference between $\mathbf{X}_A$ and $\mathbf{X}_B$ prior to standardisation. Therefore, these two quantities have to be compared, not the difference between the best *RMSEP* value for the predictions and the *RMSEP* using zero components as is often done in regular applications of PLS (see validation, Chapter 13, Figure 13.2).

The alternative univariate approach models each single wavelength separately. The method is similar to the one described above, but instead of using the whole spectrum on each side, each variable in $\mathbf{X}_A$ and $\mathbf{X}_B$ is treated separately. A new transformation $b_k$ is computed for each variable $k$. In effect, the matrix **B** in (17.1) is constrained to be diagonal, the $b_k$s being the diagonal elements.

Direct standardisation is a simple approach, but it has some drawbacks. In the multivariate case (formula 17.1), all information in instrument B is used to predict all information in A. This may be problematic if differences between the instruments are local in the wavelength domain. The univariate version on the other hand relates each individual wavelength in B to the corresponding wavelength in A. This means that it is best suited for intensity differences, not wavelength shift problems.

> Note that if PLS is used on centred data, an intercept is implicitly a part of the regression model.

*17.2.1.2 Piecewise direct standardisation (PDS)*

Piecewise direct standardisation (PDS) is a compromise between the two extremes of using either the whole spectrum or each single wavelength separately as was described above. The PDS approach was developed by Wang *et al.* (1991), Wang and Kowalski (1992) and Wang *et al.* (1993).

The method uses the same model structure and the same modelling tools as above. The only difference is that the data matrices are obtained locally in the wavelength domain. The method is based on the idea that a spectral value at wavelength $k$ for one instrument ($\underline{A}$), $x_{kA}$, can be modelled well by spectral values from a small wavelength region (window) around the same wavelength, $\mathbf{X}_{kB} = [\mathbf{x}_{(k-m)B}, ..., \mathbf{x}_{kB}, ..., \mathbf{x}_{(k+m)B}]$, in the other instrument ($\underline{B}$). The window size is defined as the number of wavelengths in the window region (window size = $2m + 1$). The wavelength $k$ is the centre of the window.

The absorbance value at wavelength $k$ for instrument $\underline{A}$ is then regressed linearly onto all wavelengths in the window centred at wavelength $k$ for the instrument $\underline{B}$. The regression model can be written as

$$\mathbf{x}_{kA} = \mathbf{X}_{kB}\mathbf{b}_k + \mathbf{e}_k \qquad (17.3)$$

Here $k$ denotes that the data and model refer to a local region centred at $k$. Note that $\mathbf{x}_{kA}$ is now a vector corresponding to the measurements for variable $k$ in the "master" instrument $\underline{A}$. Again, an intercept may be added if instruments differ in general baseline level.

As before PCR or PLS can be used for estimating regression coefficients. This results in a set of regression coefficients, $\hat{\mathbf{b}}_k$, one for each local window. If we combine the different $\hat{\mathbf{b}}_k$ vectors into a matrix $\hat{\mathbf{B}}$ we obtain a banded diagonal transformation matrix [see de Noord (1994)], intermediate in complexity between the general $\mathbf{B}$ of 17.1 and the diagonal $\mathbf{B}$ of the univariate approach.

**17.2 Different methods for spectral transfer**

Since the PDS method works locally but not just for one wavelength at a time it can handle both wavelength shift and intensity problems.

### 17.2.1.3 Selection of standard samples

The direct standardisation methods require that a number of standard samples are available for estimation of regression parameters. Selecting such samples can be done in various ways. Important ideas, principles and strategies can be found in Chapter 15, Næs (1987), Wang *et al.* (1991) and in Kennard and Stone (1969). The effect of using different standard sample sets is studied by, for example, Bouveresse *et al.* (1994).

Note that in many cases, using standard samples for calibration transfer can be quite cumbersome. For instance, food samples have a very limited lifetime and transporting such samples between instruments may be difficult. Note also that two instruments are not necessarily different in the same way for different types of products. In such cases, the instrument standardisation must be done for each product type separately. This may be very time-consuming.

### 17.2.2 Spectral transfer without the use of standard samples

One way of handling the standard sample problem is to look for a data pre-treatment that will remove inter-instrument variations in the spectra whilst preserving the signal. In Chapter 10 were described a number of pre-treatment methods that are designed to remove variations in the spectra due to differences in light scatter. In so far as the inter-instrument variability resembles scattering variability these pre-treatments will tend to remove it. However, some inter-instrument variations will not be removed by these treatments. Because they are applied globally, i.e. to the whole spectrum, they have little chance of removing variations that only affect part of the wavelength range. Such local variations may be caused by wavelength or detector differences between instruments in one region of the spectrum.

The method described by Blank *et al.* (1996) uses a local version of MSC, in which each spectral point undergoes an individual correction, using MSC on a spectral window centred on the point in question. Such a local version of MSC was proposed by Isaksson and Kowalski (1993) and has been described in Chapter 10. The method works as follows.

Each spectrum obtained from the slave instrument is compared with a target spectrum from the master instrument. An obvious choice for this target spectrum is the mean spectrum of the calibration set, though it could be the spectrum of a typical sample. Figure 17.1 shows two NIR spectra, digitised at 2 nm intervals. To correct the slave measurement at 1520 nm we might take a window from 1470 nm to 1570 nm, i.e. 50 nm on either side of the point to be corrected, plot the 51 slave measurements in this window against the 51 master measurements at the same wavelengths and fit a straight line to the plot, as in Figure 17.2. The line is fitted by least squares with slave as $y$ and master (average) as $x$. The fitted straight line is used to find the corrected version of the slave measurement, as shown in the enlargement in Figure 17.3 of the relevant portion of the plot. The equations for all this are the same as those used for MSC (see Chapter 10). Now to correct the measurement at 1522 nm we shift the window along by 2 nm and repeat the procedure. Sliding the window along the whole spectrum we can correct each wavelength in turn, except for a group of wavelengths at each end of the spectrum that do not have enough points on one side to fill the window. These can either be lost, if the window is narrow enough for this not to matter, or

**17.2 Different methods for spectral transfer**

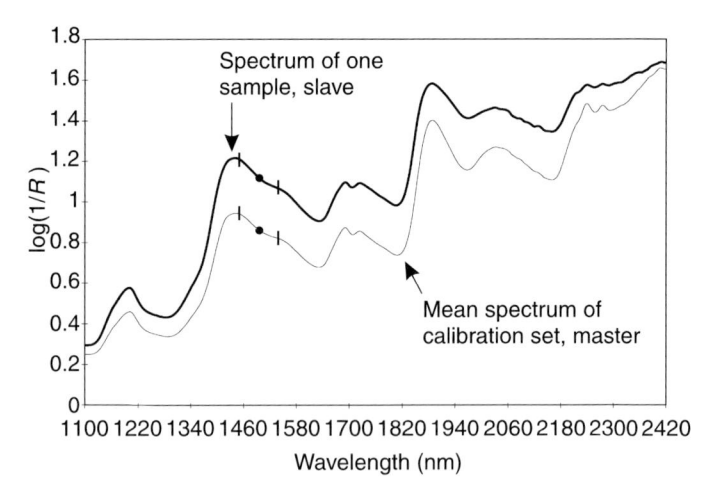

**Figure 17.1.** Mean spectrum of calibration set on master instrument (bottom) and spectrum of one sample on slave (top). The • shows 1520 nm, and the interval on either side of it indicates the window used to correct this wavelength.

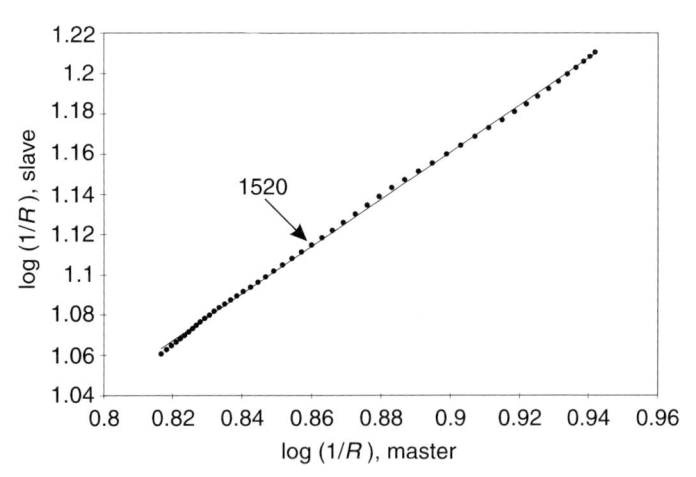

**Figure 17.2.** Slave spectrum versus master spectrum for the 51 points from 1470 to 1570 nm. The straight line has been fitted by least squares.

corrected in some other way, e.g. by a single application of MSC to the end segment.

The tricky part is getting the width of the window right. If the window is too wide the correction

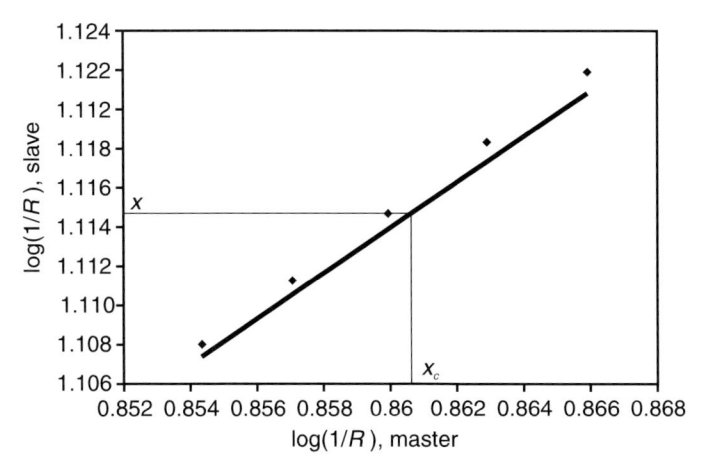

**Figure 17.3.** Enlargement of the portion of Figure 17.2 around the 26th point, at 1520 nm, showing the derivation of the corrected measurement at this wavelength. $x$ is the original measurement on the slave, $x_c$ is the corrected version.

procedure will not remove local variations due to the instrument. If it is too narrow all the information in the slave spectrum will be removed. For example, a three-point window would lead to the corrected slave spectrum being virtually indistinguishable from the target spectrum. The optimal width will depend on the size and shape of the peaks used by the calibration and on the type of inter-instrument variations encountered, and thus will be application specific.

To tune the window width we need to be in a position to assess the transferability of the calibration for various choices. We might measure some samples on master and slave, or we might just use a validation set with known chemistry measured on the slave to do this. Then the transferability of the calibration can be assessed for a range of windows, and the optimum chosen. This exercise only needs to be carried out once per calibration, not necessarily once per instrument if several instruments are to be standardised to

**17.2 Different methods for spectral transfer**

the same master, since the window may be optimised once and for all.

Thinking of this correction procedure as a pre-processing treatment, it seems logical to apply the same treatment to the calibration set before calibrating. However, there is no necessity to do this, and it is perfectly feasible to combine the correction as described with some other data treatment, second derivative for example. The point of the correction is to remove inter-instrument differences, not to remove scatter effects for a given instrument.

The original paper [Blank *et al.* (1996)] reported some promising results with this technique. The applications so far, however, seem to have been to pharmaceuticals and polymers, where the spectra tend to be more simple spectra with well-defined peaks, than they are in, say, food or agricultural applications. In the future it will be very interesting to see how the method performs in such more complicated examples.

## 17.3 Making the calibration robust

### 17.3.1 Introducing instrument variability into the calibration

The basis for this method is that all possible instrument settings/variables and other phenomena that can be changed (for instance, temperature) are varied during calibration of the instrument. In this way one hopes that instrument drifts and differences are calibrated for in such a way that small changes during regular operation will already be accounted for. Sometimes data from a number of instruments may be pooled to make a calibration set, forcing the calibration to account for the differences between these instruments.

The technique is interesting, but cannot be used to account for unforeseen effects. If instrument changes/differences are large, prediction ability can also be reduced by such a strategy.

### 17.3.2 Rugged calibration methods

This approach is due to Gemperline (1997) and is closely related to the one above. For the so-called rugged calibration approach a number of calibration models are developed, either as individually trained neural networks or as PLS models with different factors. Then each of these methods is tried on a test data set, which is a modified version of a real data set. The modification is obtained by manipulations inspired by what is believed to be typical instrument differences or drift. Gemperline proposes a number of such transformations. The calibration method which performs best on the modified test set is selected as the most rugged calibration.

### 17.3.3 Selecting robust wavelengths

Mark and Workman (1988) proposed and tested a method for selecting robust wavelengths to be used in multiple linear regression (MLR). The variable selection method starts by finding wavelengths which are invariant to spectral shift. These wavelengths are dubbed "isonumeric". The final calibration wavelengths are then chosen from among the isonumeric wavelengths in the calibration spectra. This method provides robust, transferable models and requires no transfer samples. The only samples which need to be run on the second instrument are the validation samples needed to test the transfer.

## 17.4 Calibration transfer by correcting for bias and slope

This method has been used since NIR started as a commercial technology. The method is attractive because of its simplicity, but can only be used for particular types of instrument differences that cause a change in bias or slope.

The method requires that a number of samples with known *y*-value are measured on the <u>B</u> instrument. Alternatively, these known *y*-values can be predictions from the master instrument <u>A</u>. If we denote the predicted values from the calibration equation used on the <u>B</u> instrument by $\hat{\mathbf{y}}_B$, the bias and slope correction can be obtained by regressing the known **y** values onto $\hat{\mathbf{y}}_B$ using the model:

$$\mathbf{y} = a\mathbf{1} + b\hat{\mathbf{y}}_B + \mathbf{e} \qquad (17.4)$$

Here **1** is a vector of ones and **e** is the vector of residuals.

After the model parameters *a* and *b* have been estimated by least squares, the $\hat{\mathbf{y}}_B$ for a new sample is replaced by the corrected value

$$\hat{y}_{i,B(A)} = \hat{a} + \hat{b}\hat{y}_{i,B} \qquad (17.5)$$

where *i* now denotes the index of the new sample (see Figure 17.4).

Note that the need for this approach can be checked by simple plotting of **y** vs $\hat{\mathbf{y}}_B$. If the points follow a reasonably linear relationship which is clearly skew compared to the 45° line, a bias and slope correction may be appropriate. The corrected predictor should always be validated to make sure that the new method is actually better than before correction. We also refer to the warnings given in Chapter 16.

# Standardisation of instruments

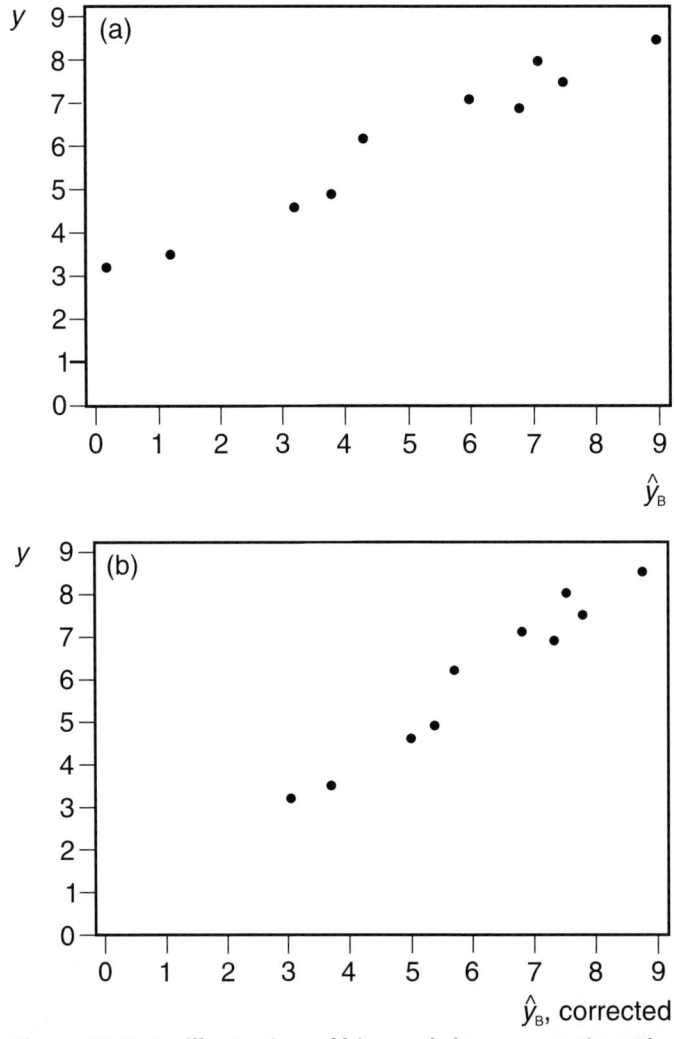

Figure 17.4. An illustration of bias and slope correction. The data are artificially generated for illustration purposes. In (a) the predicted y based on instrument <u>B</u> is plotted versus the true y values. The slope is equal to 0.64 and the intercept is 2.88. After bias and slope correction the corrected prediction based on instrument <u>B</u> is plotted against the true y [in (b)]. In this case the slope is 1 and intercept is 0. The prediction error is as we can see substantially reduced. The squared correlation coefficient is, however, the same and equal to 95.9% in both cases, showing a shortcoming of the correlation coefficient (the correlation can be large even for strongly biased solutions) for checking the validity of a calibration equation (see also Chapter 13).

## 17.4 Calibration transfer by correcting for bias and slope

# 18 Qualitative analysis/ classification

## 18.1 Supervised and unsupervised classification

Statistical classification has a number of interesting applications in chemistry. For NIR data in particular it has been used in a number of scientific publications and practical applications.

There is an important distinction between two different types of classification: so-called unsupervised and supervised classification. The first of these usually goes under the name of cluster analysis and relates to situations with little or no prior information about group structures in the data. The goal of the techniques in this class of methods is to find or identify tendencies of samples to cluster in subgroups without using any type of prior information. This is a type of analysis that is often very useful at an early stage of an investigation, to explore, for example, whether there may be samples from different subpopulations in the data set, for instance different varieties of a grain or samples from different locations. In this sense, cluster analysis has similarities with the problem of identifying outliers in a data set (see Chapter 14).

Cluster analysis can be performed using very simple visual techniques such as PCA, but it can also be done more formally, for instance by one of the hierarchical methods. These are techniques that use distances between the objects to identify samples that are close to each other. The hierarchical methods lead to so-called dendrograms, which are visual aids for deciding when to stop a clustering process (see Section 18.9.2).

**AA**
CVA: canonical variate analysis
DASCO: discriminant analysis with shrunk covariance matrices
KNN: K nearest neighbour
LDA: linear discriminant analysis
LDF: linear discriminant function
LWR: locally weighted regression
NIR: near infrared
PCA: principal component analysis
PCR: principal component regression
PLS: partial least squares
QDA: quadratic discriminant analysis
RDA: regularised discriminant analysis
*RMSEP*: root mean square error of prediction
SIMCA: soft independent modelling of class analogies
SSP: sum of squares and products

The other type of classification, supervised classification, is also known under the name of discriminant analysis. This is a class of methods primarily used to build classification rules for a number of pre-specified subgroups. These rules are later used for allocating new and unknown samples to the most probable subgroup. Another important application of discriminant analysis is to help in interpreting differences between groups of samples. Discriminant analysis can be looked upon as a kind of qualitative calibration, where the quantity to be calibrated for is not a continuous measurement value, but a categorical group variable. Discriminant analysis can be done in many different ways as will be described in Sections 18.3, 18.4, 18.5 and 18.6. Some of these methods are quite model oriented, while others are very flexible and can be used regardless of structure of the subgroups.

Material in several earlier chapters is also relevant for classification. For instance, Chapters 4, 5, 6, 10, 13, 14, 15 and 17 treat problems related to basic areas such as collinearity, data compression, scatter correction, validation, sample selection, outliers and spectral correction which are all as important for this area as they are for calibration.

In the following we will first describe discriminant analysis. Textbooks of general interest are McLachlan (1992), Ripley (1996) and Mardia *et al.* (1979).

## 18.2 Introduction to discriminant analysis

This chapter is an introduction to the basics of discriminant analysis. In order to be able to draw a picture we start with the simplest case; two groups and two variables $x_1$ and $x_2$, for instance log(1/R) at two carefully chosen NIR wavelengths (Figure 18.1).

# Qualitative analysis/classification

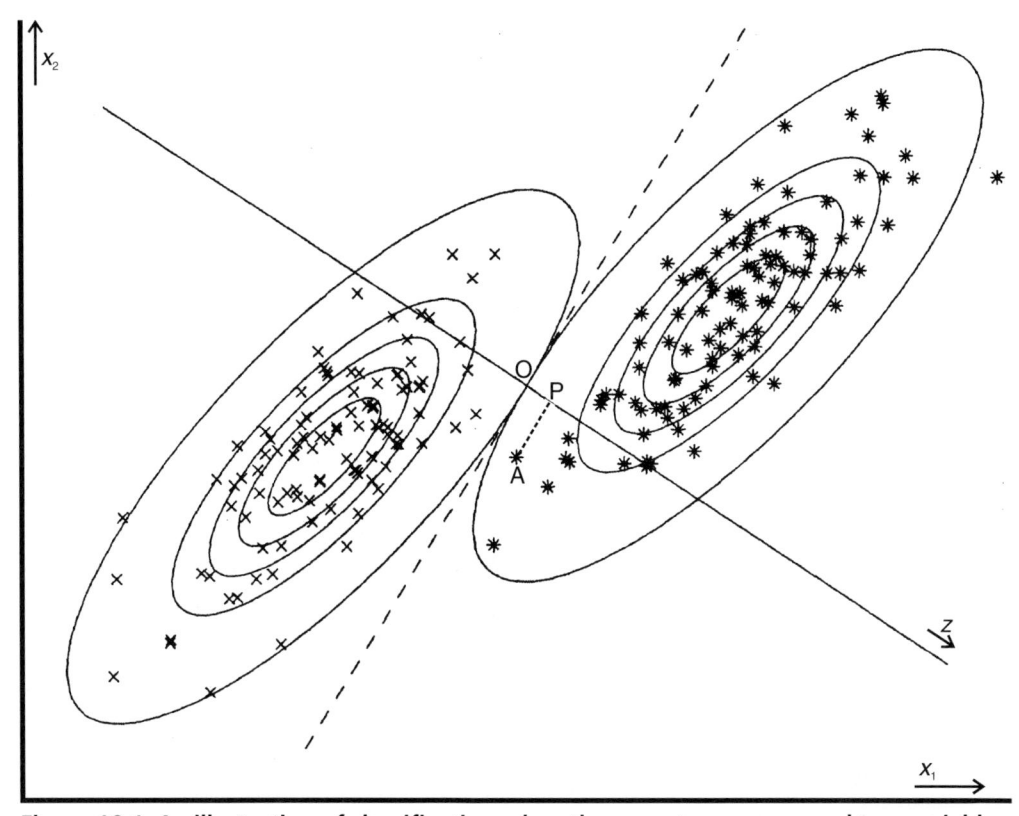

**Figure 18.1.** An illustration of classification when there are two groups and two variables. The ellipses represent probability contours of the underlying probability distribution. The dashed line indicates the splitting line between the two groups obtained by the LDA method for classification to be discussed below. The solid line indicates the direction that distinguishes best between the groups, discussed in Section 18.4.

The plotted points in Figure 18.1 are training samples, with the symbols × and * denoting the group to which each sample belongs.

This (artificial) example illustrates the selectivity problem often met in classifications of this type. This is analogous to the selectivity problem in calibration discussed in Chapter 3. Neither of the $x$ variables is selective for the property of interest, namely which group a sample belongs to. Therefore, as in many calibration examples, a multivariate vector is needed in order to obtain a successful classification.

## 18.2 Introduction to discriminant analysis

The problem of discriminant analysis is to use the plotted data to build a classifier that can be used for unknown new samples. This is equivalent to identifying a border line between the two groups. In Figure 18.1 this may seem easy, the dashed line separates the training samples perfectly, but in practice it can be quite difficult. First of all, the boundary between the two groups may be much more irregular than the one shown in Figure 18.1. In addition, there may be several more measurement variables, making it very difficult to use visually-based techniques. A third problem is the collinearity problem, as discussed extensively earlier. The latter problem can, as for the calibration methods, make the classification rule very sensitive to over-fitting.

Some approaches to discriminant analysis assume, either explicitly or implicitly, that the variation in **x** within groups may be described by multivariate normal distributions. The ellipses in Figure 18.1 are probability contours for bivariate normal distributions. Orienteers should have no difficulty in visualising the populations: see the ellipses as height contours of a pair of identical hills rising from the page. Viewed from ground (page) level the hills have as profiles the familiar bell-shaped normal probability curves. The contours actually join points of equal probability, so what you are visualising as a hill is in fact a probability distribution. For a given pair of values of $x_1$ and $x_2$ the higher we are up the hill at that point, the more probability there is of observing that particular combination of measurements. The dashed line separating the two sets of ellipses (and the two sets of points) is where the two sets of contours have equal height.

The other constructions in Figure 18.1 will be explained in section 18.4.

Figure 18.2 shows a flowchart for discriminant analysis. Training data are required to be available for

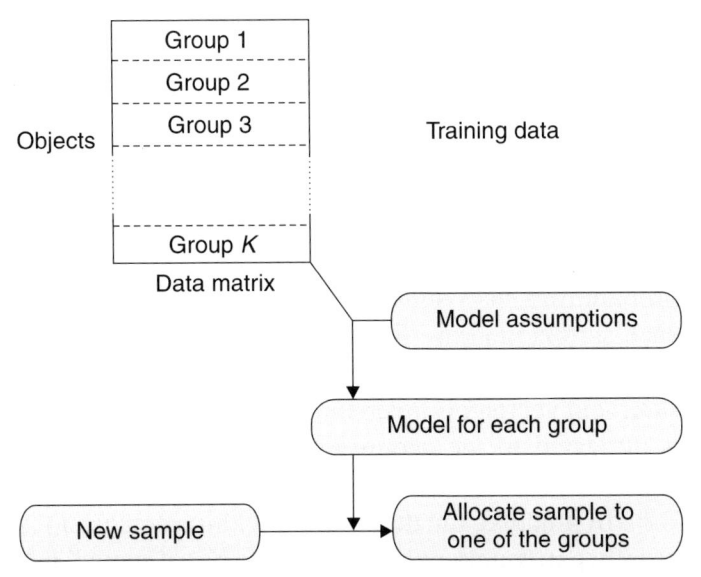

**Figure 18.2.** Discriminant analysis. Training data and model assumptions for the different groups are used for building the classifier. KNN is not covered by this flow chart (see 18.6.1).

a number of groups. These training data are used together with model assumptions to build a classifier that can be used for new samples. Note the similarity of this and the calibration procedure described extensively above.

## 18.3 Classification based on Bayes' rule

The classical way of developing a discrimination rule is by the use of the so-called Bayes' formula. In order to use this formula in the present context, one needs to assume that the probability distributions within all the groups are known and that prior probabilities $\pi_j$ for the different groups are given. These prior probabilities sum up to 100% over groups and are the prior probabilities that an unknown sample, given no knowledge of its **x**-values, belongs to the different groups. If no such information is available, one usually assumes that $\pi_j = 1/G$, where $G$ is the number of groups.

With these assumptions, using Bayes' formula to compute the so-called posterior group probabilities for each new sample is then a very easy exercise [see, for example, Mardia *et al.* (1979)]. The posterior probabilities also sum up to 100% over groups for each sample and are the probabilities that the actual sample belongs to the different groups after its measurement vector **x** is taken into account. A new sample is allocated to the group with the highest posterior probability.

In practice, the distribution within each group is usually partly unknown and needs to be estimated. For the classical methods to be discussed next, the distribution function assumed is the Gaussian or normal distribution. The training samples are in these cases used to estimate the unknown parameters in the distributions. The estimated parameters are plugged into Bayes' formula and used directly to compute estimates of the posterior probabilities.

Estimated densities based on non-parametric smoothing techniques can also be used if the data set is large. This is more complicated, but some of the standard statistical software packages can do it (see, for example, the SAS system, Raleigh, NC, USA).

It can be shown that if the assumptions about probability distributions are correct, the Bayes allocation rule is optimal with respect to the expected number of correct allocations. However, even if the assumptions are not exactly satisfied, the method is much used and can give very good and useful results.

### 18.3.1 Linear discriminant analysis (LDA)

Linear discriminant analysis (LDA) is the simplest of all possible classification methods that are based on Bayes' formula. It is based on the normal distribution assumption and the assumption that the covariance matrices of the two (or more) groups are identical. This means that the variability within each group has the same structure as is true for the two groups in Figure 18.1 for example. The only difference between groups is that they have different centres. The estimated covariance matrix for LDA is obtained by pooling covariance matrices across groups.

The means of the different groups can be estimated by the regular empirical means $\bar{x}_j$ and the pooled covariance matrix can be written as

$$\hat{\Sigma} = (N-G)^{-1} \sum_{j=1}^{G} (N_j - 1)\hat{\Sigma}_j \qquad (18.1)$$

i.e. as a weighted average of individual covariance matrices. Here the $N_j$ is the number of samples and $\hat{\Sigma}_j$ is the empirical covariance matrix for group $j$ (see for instance Appendix A) and $N$ is the sum of all the individual $N_j$s. Bayes' formula using these assumptions gives the following allocation rule for an unknown sample: allocate the unknown sample with measured vector $\mathbf{x}$ to the group $j$ with the smallest value of

$$L_j = (\mathbf{x} - \bar{\mathbf{x}}_j)' \hat{\Sigma}^{-1} (\mathbf{x} - \bar{\mathbf{x}}_j) - 2 \log_e \pi_j \qquad (18.2)$$

It can be shown that the difference $L_j - L_k$ (for groups $j$ and $k$) can be reduced to a linear function of $\mathbf{x}$, which is the source of the name LDA. For the simplest situation of two groups, the criterion leads to an allocation rule as indicated by the dashed line separating the two groups in Figure 18.1. Points to the left of the straight line will be allocated to the corresponding group and points to the right to the other. Note that if the prior probabilities are assumed identical, the second element in the sum is identical for each group. The criterion is then reduced to the squared Mahalanobis distance (see also the formula for leverage in Chapter 14).

**18.3 Classification based on Bayes' rule**

Because of the pooled covariance matrix, the LDA method is often quite stable, even for training data sets that are only moderately large. LDA is generally considered to be a robust and versatile workhorse in the area of classification. It is not always optimal, but will often give reliable results. An application of LDA in NIR spectroscopy can be found in Indahl *et al.* (1999).

### 18.3.2 Quadratic discriminant analysis (QDA)

Without the common covariance matrix assumption, Bayes' rule gives the following allocation criterion: allocate an unknown sample with measurement vector **x** to the group with the smallest value of

$$L_j = (\mathbf{x} - \bar{\mathbf{x}}_j)' \hat{\Sigma}_j^{-1} (\mathbf{x} - \bar{\mathbf{x}}_j) + \log_e |\hat{\Sigma}_j| - 2 \log_e \pi_j \quad (18.3)$$

The symbol $|\hat{\Sigma}_j|$ in (18.3) is the determinant of $\hat{\Sigma}_j$ [the determinant is a scalar function of the matrix, see for instance Healy (2000)]. The $L_j$ is a quadratic function of **x**, even for differences $L_j - L_k$, and the method is thus called quadratic discriminant analysis (QDA). The shape of the curves separating the groups of observations are curved lines in the multivariate space [see, for example, Ripley (1996) for an illustration].

As can be seen, the criterion in (18.3) is similar to LDA. The only differences are that the covariance matrices can vary from group to group and that the determinants of the covariance matrices are incorporated. Note that with equal prior probabilities the probability terms will vanish from the criterion. The same is true for covariance matrices with equal determinants which is why the determinants do not appear in (18.2).

Note that the first part of (18.3), the squared Mahalanobis distance is similar to the leverage formula used above for detecting outliers (Chapter 14).

> The determinant of $\hat{\Sigma}_j$ is equal to the product of its eigenvalues:
> (i.e. $\prod_{a=1}^{K} \hat{\lambda}_{ja}$)

**18.3 Classification based on Bayes' rule**

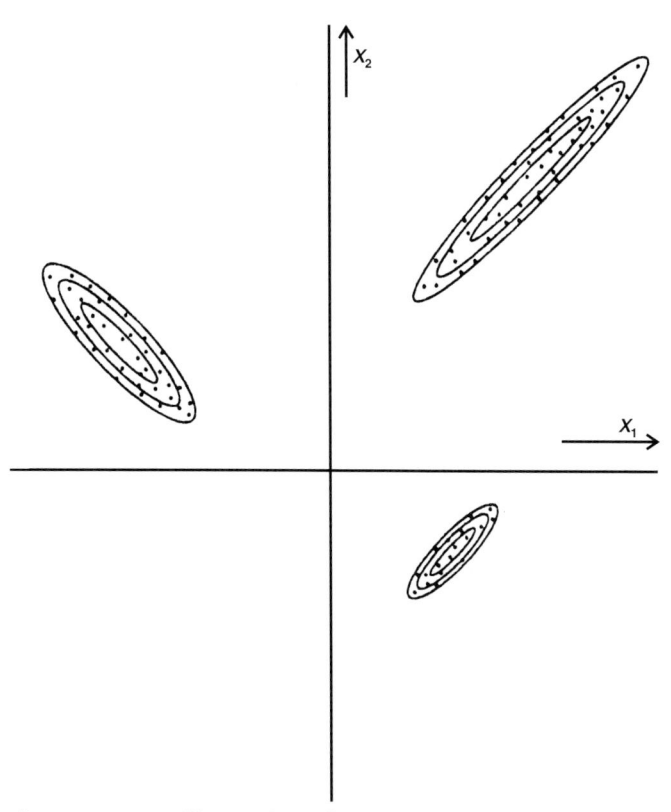

Figure 18.3. An illustration of unequal covariance matrices. As can be seen, the directions of main variability are different for the different groups. Their sizes are also different.

Mahalanobis distance is a way of measuring the distance of an observation to the centres of the groups. It uses the ellipses in the Figure 18.1 to define distance. All the points on an ellipse have the same Mahalanobis distance to the centre; the rather complicated looking formula used to calculate the Mahalanobis distance is just the equation of the ellipse.

Examples of the use of QDA in spectroscopy (based on principal components, see below) can be found in Næs and Hildrum (1997) and Indahl *et al.* (1999). An illustration of a situation with unequal covariance matrices is given in Figure 18.3. In such cases, a QDA will usually do better than a LDA unless

## 18.3 Classification based on Bayes' rule

the number of training samples is very small. In such cases, the stability of the pooled covariance matrix may become more important than the better model used for the QDA.

## 18.4 Fisher's linear discriminant analysis

In addition to providing allocation rules for unknown samples, discriminant analysis can also be used for interpreting the differences between the groups. LDA has a graphical analogue that is called Fisher's linear discriminant analysis or canonical variate analysis (CVA), which handles this problem elegantly. It seeks directions in multivariate space that separate the groups as much as possible and uses information along these directions in simple scatter plots. These plots can be used to visualise how different the groups are, to define allocation rules and also for interpreting the differences.

To be able to describe the method we need to define two matrices. The within-groups sum of squares and products (SSP) matrix is defined as

$$\mathbf{W} = \sum_{j=1}^{G} (\mathbf{X}_j - \mathbf{1}\bar{\mathbf{x}}_j')'(\mathbf{X}_j - \mathbf{1}\bar{\mathbf{x}}_j') \quad (18.4)$$

In order to make sense, this essentially assumes that all covariance matrices, as for LDA, are similar. The between-group SSP matrix, which measures the variability between the groups, is computed as

$$\mathbf{B} = \sum_{j=1}^{G} N_j (\bar{\mathbf{x}}_j - \bar{\mathbf{x}})(\bar{\mathbf{x}}_j - \bar{\mathbf{x}})' \quad (18.5)$$

Here $\bar{\mathbf{x}}$ is the average vector for the whole dataset.

Fisher's linear discriminant analysis first seeks a direction defined by the vector **a** (properly normalised), that maximises the quantity

$$\mathbf{a}'\mathbf{B}\mathbf{a} / \mathbf{a}'\mathbf{W}\mathbf{a} \quad (18.6)$$

This is equivalent to finding the direction **a** in multivariate space for which the difference between the groups' means is as large as possible compared to the within-group variance. In other words, the vector **a** defines the direction which discriminates as much as possible between all the groups. The vector **a** can technically be found as the eigenvector of

$$\mathbf{W}^{-1}\mathbf{B} \qquad (18.7)$$

which corresponds to the largest eigenvalue [see Mardia *et al.* (1979)]. This is called the first canonical variate.

For the situation with only two groups as depicted in Figure 18.1, using the direction defined by **a** for discrimination is identical to LDA as defined above. The similarity is illustrated in the following way. The solid line defined by **a** passes through the point marked by O and is at right angles to the dashed line which is the boundary produced by the LDA above. Also shown in the figure is the *projection* P of the point A on to the solid line—this is the point on the line nearest to A. We can use this projection to create a new variable $z$ by measuring the distance from O to P. Taking O as an origin, points to the left of O have negative $z$, points to the right positive $z$. Then an equivalent way of defining our allocation rule (equivalent to LDA) is to plot the point, project it on to the solid line, measure $z$ and allocate to a group according to the sign of $z$. This function is called (Fisher's) *linear discriminant function* or LDF. The value of $z$ for a particular sample is called its *score* on the LDF. Sometimes it is also called the first canonical variate. Thus in practice we do not draw the picture, we calculate the LDF score numerically and classify according to its sign.

With more than two groups we are, in general, going to need more than one discriminant axis (canonical variate) to discriminate between them. Such axes

**18.4 Fisher's linear discriminant analysis**

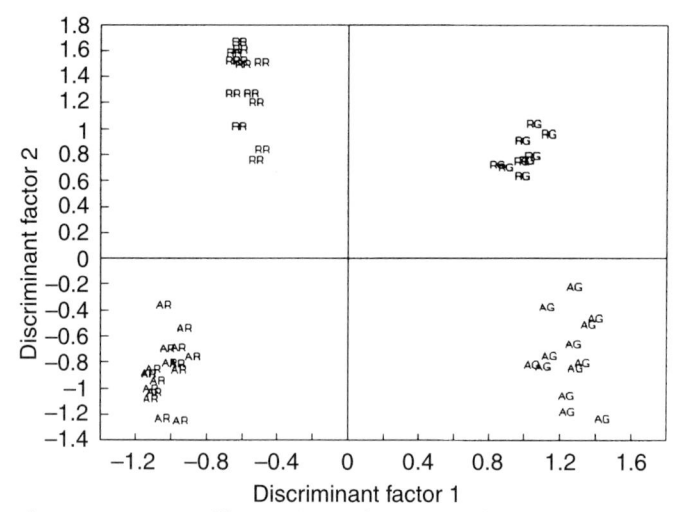

Figure 18.4. An illustration of canonical variate analysis from Downey et al. (1994). The application is discrimination between two varieties of coffee bean, Arabica (A) and Robusta (R), in two states, green (G) and roasted (R).

are defined in the same way as in equation (18.6), but in such a way that the variability along each vector is uncorrelated with variability along vectors already computed. These other axes can also be found as eigenvectors of the matrix in (18.7). In the general case it takes $G - 1$ canonical variates to discriminate between $G$ groups. Sometimes fewer will do, especially when there is some structure in the groups. Figure 18.4, taken from Downey et al. (1994), shows a case where four groups are split nicely by two, instead of the expected three, canonical variates. This is because there is a factorial structure to the groups: they involve two varieties of coffee bean, Arabica and Robusta, in two states, green and roasted. One canonical variate, rather neatly, relates to each factor.

Canonical variates are extracted in order, rather like principal components, so that the first canonical variate does the best possible job of separating all the groups, the second adds as much as possible and so on. Thus a plot of the first two canonical variates, like a plot of the first two principal components, can often tell most if not always all of the story, even when there are more than three groups. It is, however, important to stress that canonical variates and principal components are not the same things. In both cases we construct new uncorrelated variables as linear combinations of the original ones. The difference is that principal components try to represent as much as possible of the variability in the measurements, regardless of whether it is within- or between-group variability. Canonical variates, on the other hand, try to maximise the ratio of between-group to within-group variation, i.e. to find directions in which the groups are well separated. If principal components separate groups it is by accident, not because they are trying to. Of course such "accidents" often happen, because in many examples it is the between-group variability that dominates and so it is this variability that is picked up by the principal component analysis.

Another difference worth mentioning is that directions defining the canonical variates, unlike PCs, are not orthogonal in space. The scores computed are in both cases uncorrelated, but the directions (represented by the **a** vectors) used to construct them are *not* orthogonal for canonical variate analysis. A principal component analysis can be interpreted geometrically as a rigid rotation of the axes on which the original variables were measured. Finding CVs involves not merely a rotation of the axes, but also a change in their relative orientation. Some of the original right angles between axes will be reduced, others opened up.

In addition to plotting the scores, one can also plot the loadings, **a**, for the canonical variates. These indicate, as for PCA, how the discriminant directions relate to the original variables. This is done in, for instance, Ellekjær *et al.* (1993), which is a study of the effect of sodium chloride in NIR spectroscopy. In par-

**18.4** Fisher's linear discriminant analysis

ticular it was investigated whether NIR could be used to distinguish between three concentration levels of sodium chloride. The spectral pattern for the first canonical variate was used to determine which of the variables contributed most strongly to discrimination between the concentration levels. It was found that discrimination was closely related to systematic changes in the water band in the region 950–970 nm. Since the spectra were highly collinear for these data and since the canonical variates can be quite sensitive to such effects, a PCA was performed before the canonical variates were computed (see below).

## 18.5 The multicollinearity problem in classification

In most NIR and other spectroscopic applications we are not in the happy position of having many more training samples than variables. In fact the reverse is usually true. It is then not possible to calculate the inverse of the covariance matrices involved. This is the exact collinearity problem as also described above in Section 4.1. This problem makes it impossible to use the standard methods directly.

Even if it is mathematically possible to compute the exact solution, there may be situations where the classification rule, in the same way as for the calibration situation, can become extremely unstable if standard methods are applied.

A reason for this instability can easily be spotted if one looks at the LDA and QDA criteria above. Both contain the inverse covariance matrix of the variability within the groups, which can be written as

$$\hat{\Sigma}_j^{-1} = \sum_{a=1}^{K} \hat{\mathbf{p}}_{ja}\hat{\mathbf{p}}_{ja}'/\hat{\lambda}_{ja} \qquad (18.8)$$

where the $\hat{\mathbf{p}}_{ja}$s are the eigenvectors and the $\hat{\lambda}_{ja}$s are the eigenvalues of $\hat{\Sigma}_j$. If some of the eigenvalues are very small, as they will be in a situation of near collinearity (Section 4.1), even a small inaccuracy can cause a dramatic effect on the inverse covariance matrix. This can sometimes lead to very unstable classification rules, with poor classification performance.

Therefore, a number of alternative techniques to the classical LDA and QDA have been developed. Most of the methods have a very strong relationship to LDA and QDA and are based on similar ideas to the methods used for solving the collinearity problem in calibration [see Næs and Indahl (1998) for an overview]. In the following, we describe some of the most used methods and also illustrate their use through applications.

### 18.5.1 Classification based on principal components

Using principal components either for each of the groups separately or for the whole data set has become very popular for solving the collinearity problem. The simplest approach is probably to reduce the data first to a small number of principal components and use these scores as input to the discriminant analysis. This can be done for any of the methods discussed above (LDA, QDA or Fisher's linear discriminant analysis). The attractiveness of this approach lies in its simplicity, but also in the fact that it is a very useful approach. If directions with small variability within the different subgroups represent important directions for discrimination, these directions are by this approach turned into stable directions with enough variability to make them useful and reliable for discrimination. In some applications it has been found to be useful to select components not only according to the size of the eigenvalue, but also according to the discrimination power [see, for example, Downey *et al.* (1994)]. The difference between the two strategies is the same as between the strategies for selecting eigenvectors in PCR (see section 5.3.)

This approach was tested on the same data as used in Figure 3.3 to illustrate the need for multivariate methods. The data consist of NIR measurements of both pork (54 samples) and beef (49 samples). One of the things one may be interested in is to see whether it is possible to use NIR to distinguish the two animal species from each other. We computed the principal components of the spectral data and submitted the, somewhat arbitrarily chosen, nine first components to QDA and LDA. We used cross-validation for validating the performance of the classifier. In this simple illustration we used the principal components from the whole data set through the whole procedure, but in general it is recommended that the PCA is also subjected to cross-validation. It turned out that for both methods only one sample was allocated to the wrong group, one of the pork samples was classified as a beef sample. In other words, the error rate was less than 1%. We also tested the two methods using only four principal components. In this case we obtained 11 wrong allocations as measured by cross-validation, two pork samples were classified as beef and eight beef samples as pork. This shows that four principal components was not enough to obtain a good classification rule. The method was not optimised further to obtain the best possible results.

This same approach was also used in, for instance, Ellekjær *et al.* (1992). The situation was one of using NIR spectroscopy for interpreting differences between sausage products with different types of fat substitutes. Næs and Hildrum (1997) used the same approach for discriminating between tender and tough meat samples using NIR.

Another way of using principal components in classification is to perform PCA for each of the group covariance matrices separately (or for the pooled variance matrix as used in LDA) and then to replace the original matrices by truncated versions only based on the first few principal components. One then ends up with the following classification rule:

$$\hat{L}_j^{PCR} = \sum_{a=1}^{A} (\hat{t}_{ja})^2 / \hat{\lambda}_{ja}$$

$$+ \log_e \prod_{a=1}^{A} \hat{\lambda}_{ja} - 2\log_e \pi_j \qquad (18.9)$$

Note that the smallest eigenvalues are eliminated from the determinant as well.

The $\hat{t}_{ja}$ is the score along eigenvector $a$ in group $j$. All other quantities are defined as above. This criterion is used in the same way as the original method above. If the classification information is in the directions with main variability within groups, this method is usually better than the original QDA/LDA. If, however, the important information is along the smaller eigenvalue directions this method will not work (see Figure 18.5).

In such cases, the eigenvectors with small variance must be treated differently. One suggestion is to use the Euclidean distance instead of the Mahalanobis distance as used for LDA/QDA. The Euclidean distance does not divide the different directions by their corresponding eigenvalues. In this way the directions are used, but will not be so strongly influenced by the collinearity. Two established and well-known methods, SIMCA and DASCO, use such an approach.

### 18.5.2  SIMCA

SIMCA is an acronym for "Soft Independent Modelling of Class Analogies" [Wold (1976) and Wold and Sjøstrøm (1977)]. It works like this: take the x-data for one group in the training set, e.g. all the items of type A. Carry out a principal components analysis (PCA) on these data. Typically a very small number of PCs, say three or four, will represent most of the variability in the x-data. Decide on an appropri-

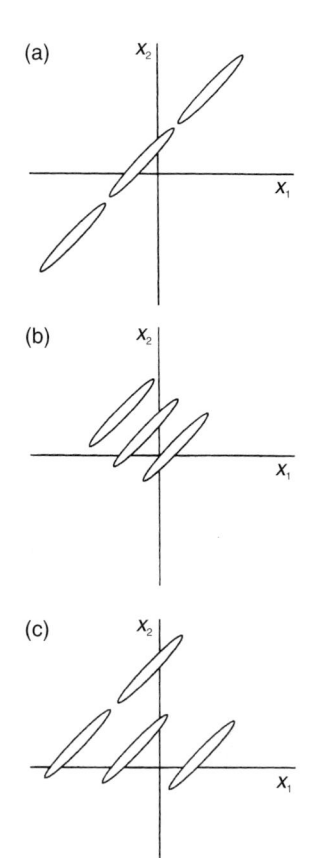

Figure 18.5. An illustration of different situations in classification. (a) depicts a situation where the discrimination power is along the direction with main within-group variability. (b) illustrates a situation where the direction of discrimination is along the direction with little within-group variability. (c) is a mixture of these two situations. Reproduced with permission from T. Næs and U.G. Indahl, J. Chemometr. 12, 205 (1998) © John Wiley & Sons Limited.

18.5 The multicollinearity problem in classification

ate number {there is scope for lots of argument about how you should do this, but cross-validation [see Chapter 13, Wold (1978)] can again be very useful} and call this the "model" for group A. Repeat this for each of the groups in the training set, to get a separate principal component model for each group.

A new unknown item is assessed against each of the groups in turn. For each group one looks at
(1) the Euclidean distance of the new item to the principal component model for that group and
(2) where the new item lies relative to the training samples within the principal component model.

Some blend of these two assessments is then used to decide whether the item could plausibly have come from this group. Note that this is exactly what was described above, namely a Mahalanobis-like distance within the principal component space and a Euclidean distance orthogonal to it.

To visualise the process, suppose that **x** has only $K = 3$ dimensions, and imagine the **x**-data for groups A and B as clouds of points in three-dimensional space (Figure 18.6). The two principal component model for Group A is a plane in the three-dimensional space chosen so that the sum of squared distances from all the A data points to the plane (measured at right angles to the plane) is as small as possible. It is the plane that gets closest (in a least squares sense) to the data. Projecting the points onto the plane gives us a two-dimensional representation of the data. The distance of a point from the plane is a residual from our model (see Chapter 14). The B data have a simpler structure, with most of the variability in one dimension, and can be represented by a one-dimensional PC model.

Now imagine a new point comes along. To assess whether it might belong to A we can project it onto the plane and see how it compares with the training samples—is it inside some box or ellipse

# Qualitative analysis/classification

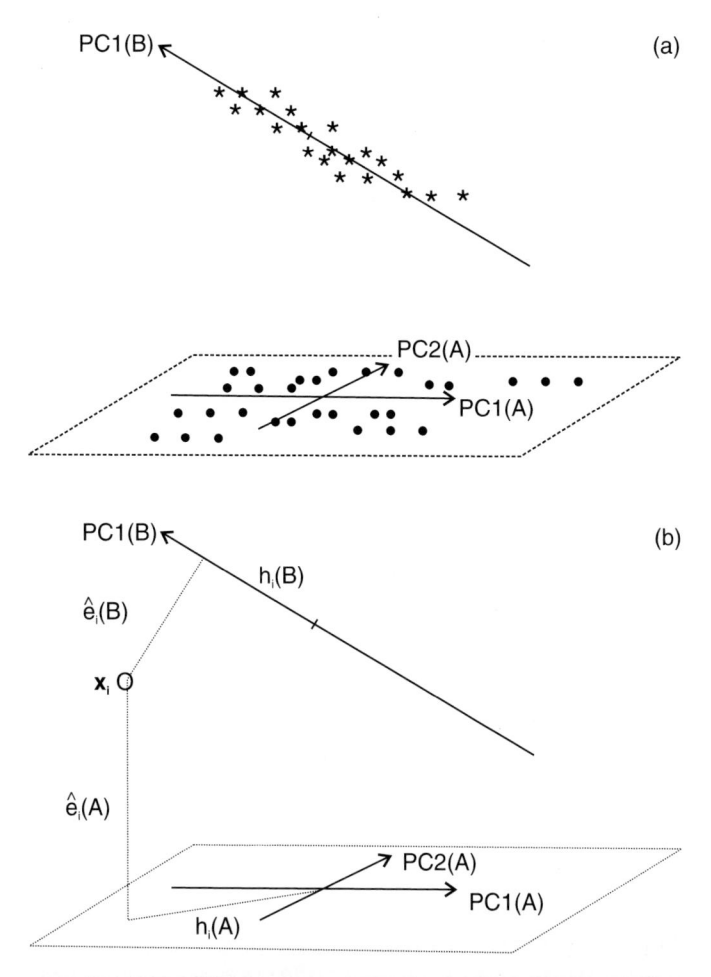

**Figure 18.6.** An illustration of SIMCA. In (a) is shown a situation with two groups modelled by a two-dimensional and a one-dimensional PC model, respectively. In (b) is shown how a new sample is compared to the two classes. As can be seen, both the distances to each PC model (the length of the es) and the distance to the centre within the class (the leverages, Mahalanobis distances, hs) are used. The smaller these numbers are for one of the groups the closer is the sample to the centre of the group.

(Mahalanobis) that describes their scatter? We can also compare its distance from the plane with those of the training samples. If it is much larger this suggests

**18.5 The multicollinearity problem in classification**

that the new point does not belong to this model. With $K = 3$ this splitting into two dimensions in the plane and one orthogonal to it achieves little. When $K = 100$ and a two-dimension principal component model the situation is, however, different. We still look individually at the two dimensions in the plane, but combine all the other 98 into a single distance to the plane. This is harder to visualise, but is computable from the PCA calculations and has the same geometric interpretation as in the case described. To assess whether the new point might belong to B we project it on to the line that is the model for group B.

An advantage of the SIMCA method is that is separates the full distance into two parts. This can be advantageous for interpretation. There is also a lot of empirical evidence for the usefulness of the method.

If SIMCA has a drawback it is that the projections are made onto different models for each group. There is no natural single picture comparable with the plot of all the data in the space of the first two canonical variates for example. This problem is solved if the approach of using PCA on the whole data set before discrimination is used (see above).

For examples of the use of SIMCA we refer to Droge and Van't Klooster (1987) and to the early paper by Wold and Sjøstrøm (1977). Many more examples have been published and can be found in the chemometrics journals.

Although at first glance SIMCA may look very different from DASCO, which is treated next, this is mainly due to the way it is presented. The two methods have strong similarities as was pointed out by Frank and Friedman (1989).

### 18.5.3 DASCO

DASCO is based on QDA as described above. The only difference between DASCO and QDA is that a modified covariance matrix is used. Instead of using the inverted sample covariance matrix, the inverse covariance matrix estimate

$$\hat{\Sigma}_j^{-1} = \sum_{a=1}^{A} \hat{\mathbf{p}}_{ja}\hat{\mathbf{p}}_{ja}' / \hat{\lambda}_{ja} + \sum_{a=A+1}^{K} \hat{\mathbf{p}}_{ja}\hat{\mathbf{p}}_{ja}' / \overline{\lambda}_j \quad (18.10)$$

is used for DASCO. Here $\overline{\lambda}_j$ is the average taken over the components $A + 1, ..., K$. Note that the first part of the equation is the same as the first $A$ elements of equation 18.8. The number of components $A$ can also be varied between the groups, but this will not be considered here. The best choice of $A$ can as usual be found by cross-validation or prediction testing (see also section 18.7).

The idea behind this choice of $\overline{\lambda}_j$ is that since the smaller eigenvalues can cause a very unstable criterion, their influence is reduced by modifying their contribution. This is done by replacing these smaller eigenvalues by their average. The contribution along the main eigenvectors is unaltered. Since the average is used along the other dimensions, the influence of the smallest eigenvalues is reduced while the influence of the moderate eigenvalues is slightly increased. Note that the relative weighting of the contributions of the two spaces is defined implicitly by the replacement of the smaller eigenvalues by their average [another weight was used in Næs and Hildrum (1997)].

The DASCO criterion for classification can then be written as

$$\hat{L}_j^{DASCO} = \sum_{a=1}^{A} (\hat{t}_{ja})^2 / \hat{\lambda}_{ja} + \sum_{a=A+1}^{K} (\hat{t}_{ja})^2 / \overline{\lambda}_j$$
$$+ \log_e |\hat{\Sigma}_j| - 2\log_e \pi_j \quad (18.11)$$

where $\hat{\Sigma}_j$ is the estimated covariance matrix given in (18.10). Essentially, this can be viewed as a sum of a Mahalanobis distance, a Euclidean distance and the other two terms used above. In order to use DASCO,

**18.5 The multicollinearity problem in classification**

one must decide how many principal components $A$ to use in the unmodified part of the covariance matrix. There is scope for lots of argument about how this should be done, but as for SIMCA it can be done by cross-validation. The DASCO method is interesting and is in our opinion not used as much as it should be.

If we compare DASCO and SIMCA we see that SIMCA neglects both the determinant term and the prior probabilities. Also, SIMCA considers the two other terms, distance to model and distance from centre within model, separately.

In the examples (simulations and real example) given in Frank and Friedman (1989) the DASCO method performs well. They also compare it with other methods such as SIMCA and regularised discriminant analysis (RDA) and the method performs favourably. Relationships between DASCO and the use of factor analysis to model the covariance structures have been pointed out by Næs and Indahl (1998).

### 18.5.4 Classification based on selected variables

An alternative strategy is to base the discriminant analysis on a small number of the original variables. Stepwise procedures for classification that select variables in sequence to optimise the discrimination are available in standard statistical packages (for instance SAS).

Another approach to this problem was suggested by Indahl *et al.* (1999). A "spectrum" of between-group variance divided by within-group variance for each variable separately was computed. The peaks of this computed spectrum were used in the classification. This procedure was tested for NIR data and gave results comparable to those obtained by using PCA on all the variables followed by the use of Bayes classification.

Which of these two strategies, using either principal components or selected variables, is to be preferred is likely to depend on the application. The arguments for and against each of them parallel those for full-spectrum analysis versus wavelength selection in quantitative analysis (see Chapter 4).

An approach that involves elements of both these strategies is to compress the data by PCA as discussed above and then select some of the principal components using a selection criterion. This means that they are not necessarily introduced in order of eigenvalue, but according to a criterion that involves classification power. This is the method adopted by Downey *et al.* (1994).

## 18.6 Alternative methods

Although the approaches described above, which are all based on or inspired by the LDA/QDA approach, are probably the most common, there are many more methods to choose from. In the following we will mention some of the most useful methods.

### 18.6.1 K nearest neighbours (KNN)

This method is, on the face of it, much simpler than any of the others that have been described above. To classify a new sample we calculate its distance from each of the samples in the training set, we find the $K$ nearest ones (typical values for $K$ are three or five, this would often be chosen to optimise performance on a test set) and classify the unknown to the group that has the most members amongst these neighbours. This is intuitively appealing, and has the advantage of making no assumptions about the shapes of the groups at all. It would still work, for example, if the configurations of the groups were as shown in Figure 18.7. With only two groups and an odd value for $K$

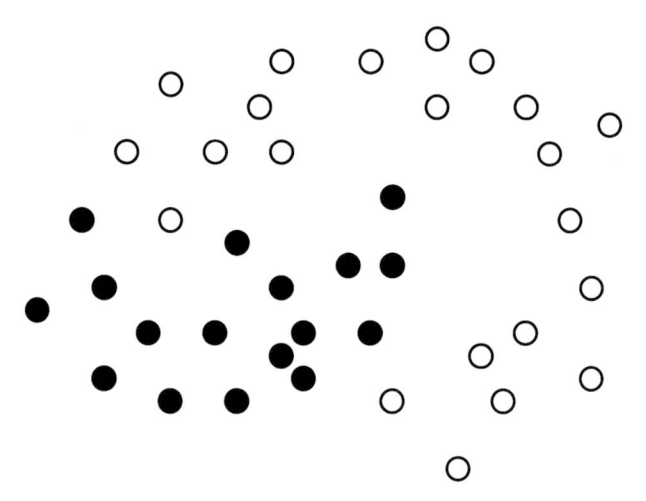

Figure 18.7. An illustration of a situation that is very difficult to handle using the LDA/QDA approach but for which KNN would work. The open and filled circles denote samples from the two groups.

there is no risk of a tied vote (which is why three and five are favoured). With more groups a tie-breaking rule is needed. Typical solutions are to use just the nearest neighbour or use the actual distances to all the neighbours in some way to decide the assignment.

Examples of the use of KNN in chemistry can be found in McElhinney *et al.* (1999) and in Thyholt and Isaksson (1997).

One feature worth remarking on is that the chance of a new item being close to a member of group A will depend, among other things, on the number of group A members in the training set. Ideally, the proportions of the groups in the training set should represent one's prior probabilities for the unknown belonging to the groups. If a rare group is heavily over represented in the training set it will tend to pick up too many assignments of unknowns.

Another drawback is that, as for all local methods, one price to be paid is the computational effort for each new sample to be classified (as for LWR in Chapter 11). Each new sample has to be compared with all the training samples, so that the data for all these have to be recalled and processed at the point of classification. If the data set is very large, this work can be substantial. Another potential drawback is that KNN may be inefficient compared with standard normal theory methods in cases where it would have been reasonable to make the usual distributional assumptions.

If the data are collinear, a possibility is to carry out a preliminary PCA on the whole training set and apply the KNN to scores on some principal components.

A general point is that the success of the KNN method will depend strongly on the way distance is measured, and there are many choices here. The most used are the Euclidean and Mahalanobis distances, as described above (Section 18.5) and in Ripley (1996).

### 18.6.2 Discrimination via regression

As stated above, discriminant analysis can be considered a qualitative calibration. Instead of calibrating for a continuous variable, one calibrates for group membership. In this section we will show how regular regression methods, as described in detail above, can be used for performing discriminant analysis.

In the case of two groups and several measurements there is a fairly obvious way to use regression methods to perform a discriminant analysis. Create a $y$-variable that has value 0 for each case in group A and 1 for each case in group B. Then use the training data to find a regression equation that predicts $y$ from the spectral measurements. Use this equation to classify unknowns according to whether the prediction is closer to 0 or 1. The choice of 0 and 1 is arbitrary, $-1$ and 1 or 0 and 100 would do just as well.

**18.6 Alternative methods**

This might seem like a different approach but in fact it is not always so. For the situation with only two groups the coefficients in the regression equation will be proportional to the weights in the canonical variate, and this is just another way of deriving the canonical variate in this two-group case. There is one slight difference: the cut-off may be in a slightly different place because of the constant term in the regression equation. If the two groups in the training data are of unequal size the regression derivation shifts the cut-off towards the smaller group, i.e. will assign more unknowns to the larger group compared with the standard discriminant analysis version.

When there are several groups, one option is to take the groups two at a time and derive a set of pairwise discriminant functions using a separate regression in each case. Another possibility is to use a genuinely multivariate regression method such as PLS2 to predict a set of 0/1 variables that describe the group memberships. In this situation one simply defines a dummy variable (0/1 variable) for each group and performs PLS2 regression of the dummy variables onto the **x**-variables. Each dummy variable is organised in such a way that it has value equal to 1 when the sample belongs to the group and 0 elsewhere. Group membership of a new unknown sample is determined by its predicted value, i.e. the sample is put in the group with the highest predicted membership value.

The collinearity is handled by using, for instance, PCR or PLS, or indeed any other favourite regression method that copes with very many collinear variables. Combinations of the PLS regression approach and LDA/QDA were attempted in Indahl *et al.* (1999). The PLS2 discriminant analysis was used to extract relevant components for classification. Then the PLS components were used for LDA and QDA in the same way as the principal components were used in section 5.1. It appeared that the classification results of this strategy were better than for the original PLS2.

Again the pork and beef data discussed in Section 18.5.1 were tested. PLS was run with a dummy $y$-variable with values 0 and 1 to identify group. Cross-validation was used for validation. With more than eight components, the classification results were good. For nine PLS components, there was, as for QDA/LDA based on nine principal components, only one erroneous classification.

We refer to Ripley (1996) for a discussion of the relationship between this approach and the other classification methods.

## 18.7 Validation of classification rules

Validation is equally important in discriminant analysis as it is in calibration. The standard measure to use instead of the *RMSEP* is the error rate or the success rate, defined as the percentage of incorrect and correct allocations, respectively. As in calibration, both prediction testing and cross-validation can be used. In some instances it may be useful to look at the error rate for each of the classes separately. There may be cases where the error rate is very small for some of the groups and much larger for others. This information may be important for interpretation and practical use of the methods. In the pork/beef example described above, cross-validation was used for estimating the error rate. In such procedures, especially when the number of samples is low, both the computation of the principal components and the classification rule should be involved in the cross-validation. This means that both the components and the classification rule should be recalculated for each segment in the validation.

The considerations about choice of training and validation samples that apply to quantitative calibration (Chapter 15) also apply to the qualitative case. In particular it is important that the training data represent all the variability within the groups. The temptation to use many replicate measurements on a small number of samples must be resisted. For instance, there is no point in training the system to discriminate between 50 scans of one particular batch from group $A$ and 50 scans of one particular batch from group $B$ if it will have to cope with considerable batch-to-batch variation when used for unknowns.

In the same way as for calibration (Section 13.7), one often wants to compare classification rules on a test set and to check for significant differences. For this situation it is possible to use the MacNemar test method [see, for example, Ripley (1996)].

## 18.8 Outliers

Outliers may be equally frequent and serious in classification as they are in calibration. Fortunately, the same methods as described in Chapter 14 can also be used here. The two most important tools, leverage and **x**-residuals are easily available for each group and can be used in the same way as described in Chapter 14.

Outliers among the unknowns can also happen. If the discrimination criterion is used uncritically, even the outliers will be put in one of the groups. It is therefore advocated that each sample "membership" is compared to the average "membership" for each of the groups. This can, for instance, be done by looking at the Mahalanobis distance. If for a particular sample, the distances to all of the G groups are very large, this sample should be identified as an outlying sample not belonging to any of the groups in the training set.

## 18.9 Cluster analysis

This section is an overview of some of the most important and most used methods for cluster analysis. We will discuss three quite different types of methods: principal component analysis, hierarchical clustering and partitioning methods. For each case we will refer to successful examples in chemistry.

We refer the reader to Kaufman and Rousseeuw (1990) for a more comprehensive and general description of cluster analysis.

### 18.9.1 Cluster analysis by the use of PCA

Principal component analysis is developed for extracting and visualising the main information in multivariate data. In previous chapters we have shown that PCA can be used for inspection of spectral patterns for improved understanding, for detecting outliers and for selecting samples for calibration. In this section we will focus on its ability to detect clusters.

If there are clearly different groups of samples in the data set, this group structure will represent a major source of variability. Using PCA, such structures will therefore show up in the score plots from the analysis. The more different the groups are, the easier they will be to see in the plots. The loadings can as usual be used for interpretation of the score axes, and loadings plots are useful for interpreting which combinations of variables characterise the different groups.

PCA is recommended for visual inspection and understanding of the group structures even if one of the other methods is used for cluster analysis. For instance, a PCA plot with a labelling of samples according to group number obtained by one of the other methods may be very useful. In general, labelling can be a very useful additional device when PCA is used for clustering. If one is anticipating a special group structure in the data set, this hypothesis can be investigated by using score plots with the correspond-

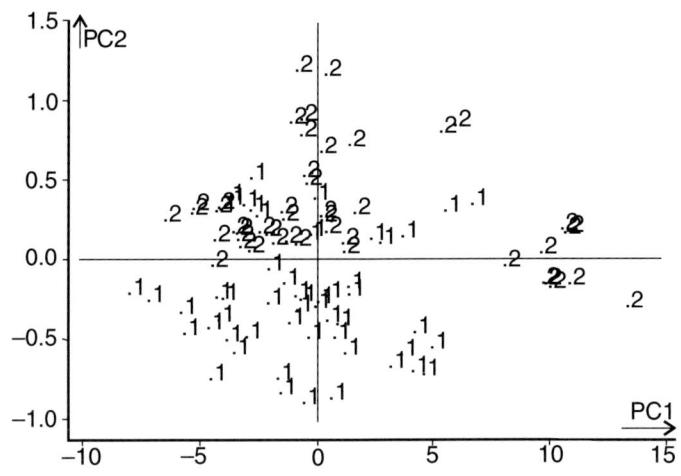

**Figure 18.8.** PCA used for cluster analysis. Based on NIR data of pork (symbol 1) and beef (symbol 2) samples.

ing labelling of samples. If samples with the same label are clearly close together in the plot this confirms the hypothesis.

Such an application of PCA is presented in Figure 18.8. This is taken from the same data set used in Figure 3.3 to illustrate the need for multivariate chemometric methods. The data consist of NIR measurements of pork and beef samples. One of the things one may be interested in is to see whether it is possible to use NIR to distinguish the two animal species from each other. The first step of an investigation of this type would typically be to run PCA on the spectral data and look at scores plots with a labelling according to the two animal species in the study. It is clear from the plot of the two first principal components in Figure 18.8 that the two groups of samples are located at different places in the plot, but that they also have a certain degree of overlap in these components. This indicates clearly a potential for distinguishing between the groups, but also that more components than

**18.9** Cluster analysis

the first two are needed in order to distinguish more clearly among them. A regular discriminant analysis based on principal components would be a natural next step in the analysis. Such results are presented in Section 18.5.

### 18.9.2 Hierarchical methods

The class of hierarchical methods can be split up into two groups, so-called *agglomerative* and *divisive* methods. The agglomerative methods start the search for clusters by treating each sample as a separate group and end up with all samples in the same large group. The divisive methods do the opposite, they start with all samples in one group and split the group into clusters in a hierarchical way. We will focus here on two much used agglomerative methods, *single linkage clustering* and *complete linkage clustering*. For more information about these and other methods, see, for example, Mardia *et al.* (1979) and Kaufman and Rousseeuw (1990).

For all hierarchical methods, a distance measure has to be defined. Obvious and much used candidates are the Euclidean and Mahalanobis distances, but several others also exist [see Mardia *et al.* (1979)]. The distances between all the samples in the data set can be presented in a distance matrix. This is a symmetric matrix with zeros on the diagonal (corresponding to the distance between a sample and itself) and all other distances in the off-diagonal cells. The distance matrix will here be denoted by

$$\mathbf{D} = \begin{pmatrix} d_{11} & d_{12} & d_{13} & \cdot & \cdot \\ d_{21} & d_{22} & d_{23} & \cdot & \cdot \\ d_{31} & d_{32} & d_{33} & \cdot & \cdot \\ \cdot & \cdot & \cdot & \cdot & \cdot \\ \cdot & \cdot & \cdot & \cdot & d_{NN} \end{pmatrix} \qquad (18.12)$$

**18.9** Cluster analysis

where $d_{ij}$ is the distance between objects $i$ and $j$. Note that $d_{ii} = 0$ and $d_{ij} = d_{ji}$ (**D** is symmetric).

The *single linkage* method starts with each sample as a separate cluster. In the first step of the algorithm, the two samples with the smallest distance value are put into the same cluster. The distance from sample $i$ to the two-sample cluster, is then defined as the distance to the sample ($k$) in the group which is closest, i.e. as

$$d = \min(d_{ik}) \qquad (18.13)$$

where the minimum is taken over the two samples ($k$) in the cluster. A new distance matrix (**D**) with one row and one column fewer than the original can be set up. In the next step one joins the two next samples with the smallest $d$ value. If the smallest $d$ value is a distance to the two-sample cluster, the sample is added to this cluster. The same procedure as above for generating distances between samples and clusters is repeated in order to obtain a new reduced distance matrix. The procedure continues until all samples belong to one large cluster, or until the desired number of clusters is obtained. At later stages of the process one also needs distances between clusters. These are defined the same way as in (18.13) as the minimum distance between a sample in one cluster and a sample in the other.

Complete linkage is an almost identical procedure. The only difference is that distances between samples and groups or between groups are defined using the maximum distance instead of the minimum as was the case for single linkage (18.13). As for single linkage, however, the cluster algorithm merges samples or groups according to closeness.

The way that the results from the two algorithms are presented and interpreted is by the use of so-called dendrograms. These are tree-like structures indicating which samples were joined into which cluster as a function of the distances between them. The plot presents distance between samples on one of the axes and sample number on the other. If there is a long gap between two distances at which samples are joined into clusters, this is a sign of a clear group structure. In other words, if one has to go much further out in space before new samples are joined into clusters, the clusters already obtained are nicely separated.

As an illustration of the use of complete linkage and the dendrogram we applied the procedure to a small data set with a clear cluster structure in it. The data are artificially generated for illustration and are presented graphically in Figure 18.9(a). As can be seen there are three clearly separated groups in it, each containing three samples. The distance measure used in the clustering is the regular Euclidean distance. The dendrogram is presented in Figure 18.9(b). The distance between samples is presented on the y-axis and sample number is on the x-axis. From the dendrogram it is clear that samples 4 and 5 are the closest, joined at a distance approximately equal to 0.8. This corresponds well to the data in Figure 18.9(a). Then sample 6 is joined with the cluster consisting of samples 4 and 5 at a distance approximately equal to 1. The next step is to join samples 7 and 8, then 1 and 2 and so on until all samples are joined at a distance of about 6.2. After three groups have been formed at a distance close to 2.0, one has to go a long way (from 2 to about 5) in the distance direction before a new joining of groups takes place. This is a clear indication that there are three clearly separated groups in the data set, which corresponds exactly to how the data were generated in Figure 18.9(a).

The two methods described, single and complete linkage, have somewhat different properties with respect to what kind of clusters they discover. Since single linkage is based on minimum of distances, it will

**18.9 Cluster analysis**

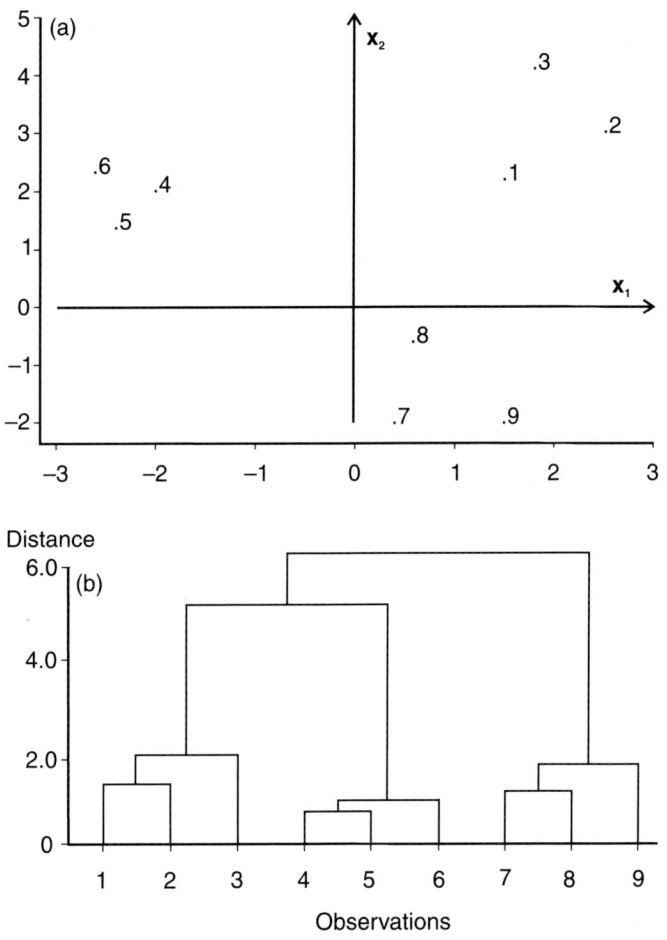

**Figure 18.9.** A graphical illustration of complete linkage. Three groups of samples were generated (a). The dendrogram of complete linkage is shown in (b).

have equally strong tendencies to allocate samples to existing groups or to join new samples into clusters. This can lead to "rod" type elongated clusters [Mardia *et al.* (1979)]. Thus, single linkage will usually not give satisfactory results if intermediate samples are positioned between clusters. Complete linkage will, on the other hand, favour joining clusters before allocating new samples to clusters, leading to more compact clusters without a chaining effect.

Note that hierarchical clustering can be useful even in situations where no clear clustering is possible. For instance, the technique for selecting calibration samples presented in Chapter 15 is such a situation. The algorithm was run until the desired number of clusters was found and then one sample was picked from each cluster.

Other methods of hierarchical cluster analysis, using both other criteria and other distance measures can be found in Mardia *et al.* (1979) and Kaufman and Rousseeuw (1990).

### 18.9.3 Partitioning methods

The hierarchical methods have the advantage that they provide a graphical aid to help in interpreting the clusters and determining how many natural clusters there are in the data. Their drawback, however, is that the cluster structure may be strongly dependent on which strategy for computing distances between clusters is used.

Another class of clustering methods, which is not based on a hierarchical search, is the partitioning methods. These are methods which optimise a criterion measuring the degree of clustering in the data. Instead of being strategy dependent, these methods are dependent on what kind of criterion is used for the optimisation.

The partitioning methods need a guess for the number of subgroups before the optimal clustering is computed. After computation of the clusters, the selected value for the number of clusters can be evaluated (see below).

If we let $d_{ij}$ denote the distance from sample $i$ to group $j$, the partitioning methods typically minimise a criterion like

$$J = \sum_{j=1}^{G} \sum_{i=1}^{N} I_{ij} d_{ij}^2 \qquad (18.14)$$

Here $I_{ij}$ is an indicator value which is 0 or 1 depending on whether sample $i$ belongs to group $j$ or not. This is natural since small distances represent a good splitting of the data set. The algorithm for optimising $J$ will then have to search for the splitting into subgroups that minimises the criterion (18.14). Different types of search algorithms have been developed for this. Testing all combinations is of course theoretically possible, but this will take a lot of time if the number of objects is large. Therefore numerical optimisation is usually needed.

Again, plotting of the clusters in a PCA plot can help in interpreting them and deciding how many to use. Cluster centres are typical representatives for the clusters and may also be used for interpretation.

### 18.9.3.1 Fuzzy clustering

Fuzzy clustering [see, for example, Bezdec (1981), Bezdec *et al.* (1981a,b)] is a partitioning method which uses a slightly different criterion than the $J$ defined above. One of the advantages of fuzzy clustering is that it provides an extra tool which can be used to interpret and determine the number of clusters, namely the matrix of membership values. In addition to the distance measure $d_{ij}$, the method uses a membership matrix consisting of values between 0 and 1. The membership matrix is here denoted by **U** and consists of $N$ rows (the number of samples) and $G$ columns (the number of groups). The elements in each line in the matrix sum up to 1 and indicate the membership values for each sample for each of the classes. The criterion to be minimised for fuzzy clustering is given by

$$J = \sum_{j=1}^{G} \sum_{i=1}^{N} u_{ij}^m d_{ij}^2 \qquad (18.15)$$

The value of $m$ must be set by the user, as described below.

The optimal solution for $J$ favours combinations of small values of $u$ with large values of $d$ and vice versa. The solution to the problem is not a crisp clustering of the samples into clusters as was the case above, but a matrix of $u$-values indicating the degree of membership of each sample to each of the subgroups. These values can be extremely useful when interpreting the clusters and when determining a useful value of $G$. Values of $u$ close to 0 or close to 1 indicate samples which are clearly allocated to one of the subgroups, while samples with values in the range close to $1/G$ are uncertain candidates. If the latter is observed for many samples in the data set, another value of $G$ should be tried. Other and more formal ways of assessing the number of clusters are discussed in Windham (1987). They can also be analysed by PCA as was proposed by Rousseeuw *et al.* (1989). A simple and useful way of plotting $u$-values in the case of three groups is shown in Figure 18.10.

The criterion $J$ has a very simple optimisation procedure. This is presented in Bezdec (1981) and can be used with several distance measures, for instance Euclidean and Mahalanobis. The two steps in the algorithm are based on an initial guess for **U**, either random or based on prior knowledge. The better the guess, the quicker the convergence.

For given $u$-values, the best possible Euclidean $d$-values are found by computing Euclidean distances relative to centres

$$\bar{\mathbf{x}}_j = \sum_{i=1}^{N} u_{ij}^m \mathbf{x}_i / \sum_{i=1}^{N} u_{ij}^m \qquad (18.16)$$

When $d$-values have been found, optimal $u$-values are determined by the formula

**18.9 Cluster analysis**

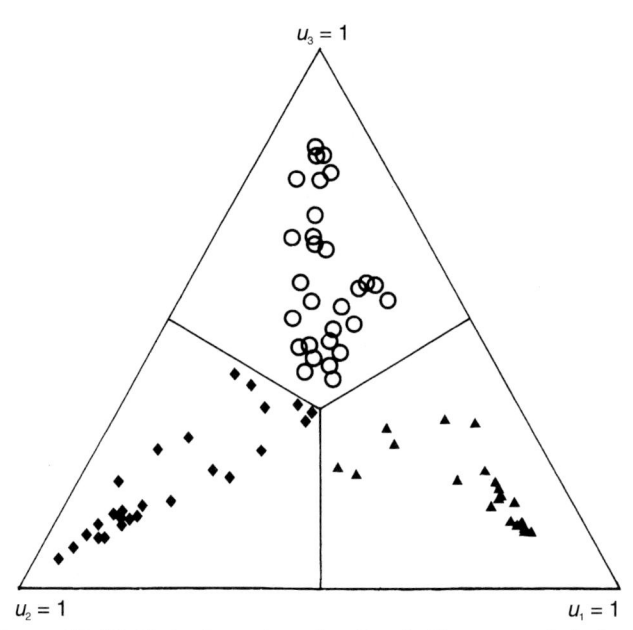

**Figure 18.10.** Plot of $u$-values for $G = 3$. The example is from the second of the applications of fuzzy clustering to the calibration problem described in section 12.2.3. The data are based on NIR measurements of pork and beef and were originally analysed in Næs and Isaksson (1991). As can be seen, some of the samples are very close to the corners of the triangle indicating that these are clear members of one of the groups. Samples close to the centre have approximately the same membership value for each of the groups. Reproduced with permission from T. Næs and T. Isaksson, *J. Chemometr.* **5**, 49 (1991) © John Wiley & Sons Limited.

$$u_{ij} = \left[ \sum_{g=1}^{G} (d_{ij}/d_{ig})^{2/(m-1)} \right]^{-1} \quad (18.17)$$

These two steps are repeated until convergence. Usually, this happens within a relatively small number of iterations.

Distances based on PCA within each cluster [Bezdec *et al.* (1981b)] have also been developed. This can be useful when the shapes of the clusters are far from circular and when the cluster structures are very different.

Fuzzy clustering has been used in many types of applications. In NIR spectroscopy it has, for instance, been used to split calibration data into subgroups with good linearity in each of the classes (see Section 12.2). For another application we refer to Gunderson *et al.* (1988).

## 18.10 The collinearity problem in clustering

As usual in chemometrics, **x**-variables are often highly collinear. This must be taken into account in the distance measure. It is most easily done by using PCA before the actual computation of distance. This is the same as truncating the regular distance. More information about truncated distances and their use can be found in Chapter 11. The fuzzy clustering method in Bezdec *et al.* (1981b) using disjoint PCA models for the different groups (as in SIMCA) can also be useful for handling the collinearity problem.

# 19 Abbreviations and symbols

## 19.1 Abbreviations

*AIC*: Akaike information criterion
ANN: artificial neural network
ANOVA: analysis of variance
CARNAC: comparison analysis using restructured near infrared and constituent data
*CN*: condition number
CR: continuum regression.
CV: cross-validation, canonical variate
CVA: canonical variate analysis
DASCO: discriminant analysis with shrunk covariance matrices
DS: direct standardisation
DWT: discrete wavelet transform
*E*: expectation
FFT: fast Fourier transform
FT: Fourier transform
GLM: generalised linear model
KNN: *K* nearest neighbours
LDA: linear discriminant analysis
LDF: linear discriminant function
LS: least squares
LWR: locally weighted regression
ML: maximum likelihood
MLR: multiple linear regression
MSC: multiplicative scatter (or signal) correction
*MSE*: mean square error
*MSEBT*: bootstrap mean square error
*MSEP*: mean square error of prediction
NIR: near infrared
OS: optimised scaling
OSC: orthogonal signal correction

PCA: principal component analysis
PCR: principal component regression
PCS: principal component shrinkage
PDS: piecewise direct standardisation
PET: polyethylene terephthalate
PLC-MC: path length correction with chemical modelling
PLS: partial least squares (regression)
PLS2: partial least squares with multiple $y$-variables
PMSC: piecewise multiplicative scatter correction
PP: projection pursuit
QDA: quadratic discriminant analysis
*RAP*: relative ability of prediction
RDA: regularised discriminant analysis.
*RMSE*: root mean square error
*RMSEBT*: bootstrap root mean square error
*RMSEC*: root mean square error of calibration
*RMSECV*: root mean square error of cross-validation
*RMSEP*: root mean square error of prediction
RR: ridge regression
*RSS*: residual sum of squares
*SEP*: standard error of prediction
SIMCA: soft independent modelling of class analogies
SMLR: stepwise multiple linear regression
SNV: standard normal variate
SPC: statistical process control
SSP: sum of squares and products
SVD: singular value decomposition
*VIF*: variance inflation factor
WLS: weighted least squares

## 19.2 Important symbols

$A$: absorbance, number of components in PCR and PLS regression

**B**, **b**, $b_0$: regression coefficients

$C_p$: Mallows $C_p$ for determining model size

$d$: distance measure used in classification
$D_i$: Cook's influence measure
**E,e**: **X**-residual matrix, vector
**F, f**: $y$-residual
$G$: number of subgroups
$K$: number of $x$-variables
$\kappa$ (kappa): condition number
$\hat{\lambda}$ (lambda hat): eigenvalue
$N$: number of objects
$\pi$ (pi): prior probabilities in discriminant analysis
**P**: **X**-loadings, eigenvectors of covariance matrix
$\sigma$ (sigma): standard deviation
$\sigma^2$: variance
$\Sigma$ (capital sigma): covariance matrix
**q**: $y$-loadings, regression coefficients for scores
$r$: correlation coefficient
$R^2$: squared multiple correlation coefficient
**U**: matrix of membership values in fuzzy clustering
**W**: loading weights in PLS
**X**: independent, explanatory or predictor variables
**Y**: dependent variable, reference value
$\hat{y}$ ($y$-hat): predicted $y$-value

**19.2 Important symbols**

# 20 References

Aastveit, A.H. and Marum, P. (1991). On the Effect of Calibration and the Accuracy of NIR Spectroscopy with High Levels of Noise in the Reference Values. *Appl. Spectrosc.* **45,** 109–115.

Aastveit, A.H. and Marum, P. (1993). Near-Infrared Reflectance Spectroscopy: Different Strategies for Local Calibration in Analysis of Forage Quality. *Appl. Spectrosc.* **47,** 463–469.

Adhihetty, I.S., McGuire, J.A., Wanganeerat, B., Niemczyk, T.M. and Haaland, D.H. (1991). Achieving Transferable Multivariate Spectral Calibration Models: Demonstration With Infrared Spectra of Thin-Film Dielectrics On Silicon. *Anal. Chem.* **63,** 2329–2338.

Akaike, H. (1974). A New Look at the Statistical Model Identification. *IEEE Transactions on Automatic Control* **AC-19,** 716–723.

Almøy, T. (1996). A Simulation Study on Comparison of Prediction Methods when only a Few Components are Relevant. *Comp. Statist. Data Anal.* **21,** 87–107.

Almøy, T. and Haugland, E. (1994). Calibration Methods for NIRS Instruments: A Theoretical Evalution and Comparisons by Data Splitting and Simulations. *Appl. Spectrosc.* **48,** 327–332.

Alsberg, B.K., Woodward, A.M. and Kell, D.B. (1997). An Introduction to Wavelet Transforms for

Chemometricians: a Time-Frequency Approach. *Chemometr. Intell. Lab.* **37,** 215–239.

Anderson, T.W. (1971). *The Statistical Analysis of Time Series*. John Wiley & Sons, New York, USA.

Barnes, R.J., Dhanoa, M.S. and Lister, S.J. (1989). Standard Normal Variate Transformation and Detrending of Near Infrared Diffuse Reflectance. *Appl. Spectrosc.* **43,** 772–777.

Barnett, V. and Lewis, T. (1978). *Outliers in Statistical Data.* John Wiley & Sons, Chichester, UK.

Barron, A.R. and Barron, R.L. (1988). Statistical Learning Networks: A Unifying View. In *Computing Science and Statistics, Proceedings of the 20$^{th}$ Symposium on the Interface*, Ed by Wegman, E. American Statistical Association, Alexandria, VA, USA, pp. 192–203.

Beebe, K.R., Pell, R.J. and Seasholtz, M.B. (1998). *Chemometrics, A Practical Guide.* John Wiley & Sons, New York, USA.

Beer, A. (1852). *Annalen der Physik und Chime* **86,** 78–88.

Belsley, D.A., Kuh, E. and Welch, R.E. (1980). *Regression Diagnostics.* John Wiley & Sons, New York, USA.

Berglund, A. and Wold, S. (1997). INLR, Implicit Non-linear Latent Variables Regression. *J. Chemometr.* **11,** 141–156.

Berzaghi, P, Shenk, J.S. and Westerhaus, M.O. (2000). LOCAL Prediction with Near Infrared Multi-

Product Databases. *J. Near Infrared Spectrosc.* **8,** 1–9.

Bezdec, J.C. (1981). *Pattern Recognition with Fuzzy Objective Function Algorithms*. Plenum, New York, USA.

Bezdec, J.C., Coray, C, Gunderson, R.W. and Watson, J. (1981a). Detection and Characterisation of Cluster Substructure. I. Linear Structure: Fuzzy c-lines. *SIAM J. Appl. Math.* **40,** 339–357.

Bezdec, J.C., Coray, C, Gunderson, R.W. and Watson, J. (1981b). Detection and Characterisation of Cluster Substructure. II. Fuzzy c-varieties and Convex Combinations Thereof. *SIAM J. Appl. Math.* **40,** 358–372.

Blank, T.B., Sum, S.T. and Brown, S.D. (1996). Transfer of Near-Infrared Multivariate Calibrations without Standards. *Anal. Chem.* **68,** 2987–2995.

Borggaard, C and Thodberg, H.H. (1992). Optimal Minimal Neural Interpretation of Spectra. *Anal. Chem.* **64,** 545–551.

Bouguer, P. (1729). *Essai d'optique sur la gradation de la lumière*. Published posthumously: Bouguer, P. and de la Caille, N.L. (1760). *Traite d'optique sur la gradation de la lumiere*, De l'imperimerie de H. L. Guerin & L. F. Delatour, Paris.

Bouveresse, E., Massart, D.L. and Dardenne, P. (1994). Calibration Transfer across Near-Infrared Spectrometric Instruments using Shenk's Algorithm: Effects of Different Standardization Samples. *Anal. Chim. Acta* **297,** 405–416.

Bouveresse, E. and Massart, D.L. (1996). Standardisation of Near-Infrared Spectrometric Instruments: A Review. *Vib. Spectrosc.* **11,** 3–15.

Box, G.E.P. and Draper, N. (1959). A Basis for the Selection of Response Surface Design. *J. Am. Statist. Assoc.* **54,** 622–653.

Box, G.E.P., Hunter, W.G. and Hunter, J.S. (1978). *Statistics for Experimenters.* John Wiley & Sons, New York, USA.

Brown, P.J. (1992). Wavelength Selection in Multicomponent Near-Infrared Calibration. *J. Chemometr.* **6,** 151–161.

Brown, P.J. (1993). *Measurement, Regression and Calibration.* Clarendon Press, Oxford, UK.

Chui, C.K. (1992). *An Introduction to Wavelets.* Academic Press, San Diego, CA, USA.

Chui, C.K., Montefusco, L. and Puccio, L. (Eds) (1994). *Wavelets: Theory, Algorithms and Applications.* Academic Press, San Diego, CA, USA.

Cleveland. W.S. and Devlin, S.J. (1988). Locally Weighted Regression: an Approach to Regression Analysis by Local Fitting. *J. Am. Statist. Assoc.* **83,** 596–610.

Cook, R.D. and Weisberg, S. (1982). *Residuals and Influence in Regression.* Chapman and Hall, London, UK.

Cooley, J.W. and Tukey, J.W. (1965). An Algorithm for the Machine Computation of Complex Fourier Series. *Math. Comput.* **19,** 297–301.

Daubechies, I. (1992). *Ten Lectures on Wavelets.* Society for Industrial and Applied Mathematics, Philadelphia, PA, USA.

Davies, A.M.C., Britcher, H.V. Franklin, J.G. Ring, S.M., Grant, A. and McClure, W.F. (1988). The Application of Fourier-Transformed NIR Spectra to Quantitative Analysis by Comparison of Similarity Indices. (CARNAC), *Mikrochim. Acta* **I,** 61–64.

Davis, L. (Ed), (1987). *Genetic Algorithms and Simulated Annealing.* Pitman, London, UK.

De Jong, S. (1995). PLS Shrinks. *J. Chemometr.* **9,** 323–326.

de Noord, O.E. (1994). Multivariate Calibration Standardization. *Chemometr. Intell. Lab.* **25,** 85–97.

Despagne, F., Walczak, B. and Massart, D.L. (1998). Transfer of Calibration of Near-Infra-Red Spectra using Neural Networks. *Appl. Spectrosc.* **52,** 732–745.

Despagne, F., Massart, D.L. and Chabot, P. (2000). Development of a Robust Calibration Model for Non-Linear In-Line Process Data. *Anal. Chem.* **72,** 1657–1665.

Donoho, D.L. and Johnstone, I.M. (1994). Ideal Spatial Adaptation by Wavelet Shrinkage. *Biometrika* **81,** 425–455.

Downey, G., Boussion, J. and Beauchene, D. (1994). Authentication of Whole and Ground Coffee Beans by Near Infrared Reflectance Spectroscopy. *J. Near Infrared Spectrosc.* **2,** 85–92.

Draper, N.R. and Smith, H. (1981). *Applied Regression Analysis.* John Wiley & Sons, New York, USA.

Droge, J.B.M. and Van't Klooster, H.A. (1987). An Evaluation of SIMCA. Part 1—The Reliability of the SIMCA Pattern Recognition Method for a Varying Number of Objects and Features. *J. Chemometr.* **1**, 221–230.

Efron, B. (1982). *The Jackknife, the Bootstrap and other Resampling Techniques.* Society for Industrial and Applied Mathematics, Philadelphia, PA, USA.

Efron, B. and Gong, G. (1983). A Leisurely Look at the Bootstrap, the Jackknife and Cross-Validation. *Am. Stat.* **37**, 36–48.

Efron, B. and Tibshirani, R.J. (1993). *An Introduction to the Bootstrap.* Chapman and Hall, New York, USA.

Ellekjær, M.R., Næs, T., Isaksson, T. and Solheim, R. (1992). Identification of Sausages With Fat-Substitutes Using Near-infrared Spectroscopy. In *Near Infra-Red Spectroscopy: Bridging the Gap between Data Analysis and NIR Applications*, Ed by Hildrum, K.I., Isaksson, T., Naes, T. and Tandberg, A. Ellis Horwood, Chichester, UK, pp. 321–326.

Ellekjær, M.R., Hildrum, K.I., Næs, T. and Isaksson, T. (1993). Determination of the Sodium Chloride Content of Sausages by Near Infrared Spectroscopy. *J. Near Infrared Spectrosc.* **1**, 65–75.

Esbensen, K.H. (2000). *Multivariate Data Analysis - In Practice. An Introduction to Multivariate Data Analysis and Experimental Design*, 4th Edn. CAMO ASA, Oslo, Norway.

Fearn, T. (1983). Misuse of Ridge Regression in the Calibration of a Near Infrared Reflectance Instument. *Appl. Statist.* **32**, 73–79.

Fearn, T. (1992). Flat or Natural? A Note on the Choice of Calibration Samples. In *Near Infra-Red Spectroscopy: Bridging the Gap between Data Analysis and NIR Applications*, Ed by Hildrum, K.I., Isaksson, T., Naes, T. and Tandberg, A. Ellis Horwood, Chichester, UK, pp. 61–66.

Fearn, T. (2000). On Orthogonal Signal Correction. *Chemometr. Intell. Lab.* **50,** 47–52.

Forbes, J.D. (1857). Further Experiments and Remarks on the Measurement of Heights by the Boiling Point of Water. *Trans. Roy. Soc. Edin.* **21,** 235–243.

Forina, M., Drava, G., Armanino, C., Boggia, R., Lanteri, S., Leardi, R., Corti, P., Giangiacomo, R., Galliena, C., Bigoni, R., Quartari, I., Serra, C., Ferri, D., Leoni, O. and Lazzeri, L. (1995). Transfer of Calibration Function in Near-Infrared Spectroscopy. *Chemometr. Intell. Lab.* **27,** 189–203.

Frank, I.E. (1987). Intermediate Least Squares Regression Method. *Chemometr. Intell. Lab.* **1,** 233–242.

Frank, I.E. (1990). A Nonlinear PLS Model. *Chemometr. Intell. Lab.* **8,** 109–119.

Frank, I.E. and Friedman, J.H. (1989). Classification: Oldtimers and Newcomers. *J. Chemometr.* **3,** 463–475.

Frank, I.E. and Friedman, J.H. (1993). A Statistical View of Some Chemometrics Regression Tools. *Technometrics* **35,** 109–135.

Friedman, J.H. and Stuetzle, W. (1981). Projection Pursuit Regression. *J. Am. Statist. Assoc.* **76,** 817–823.

Geladi, P., McDougall, D. and Martens, H. (1985). Linearisation and Scatter Correction for Near Infrared Reflectance Spectra of Meat. *Appl. Spectrosc.* **39,** 491–500.

Gemperline, P.J. (1997). Rugged Spectroscopic Calibration for Process Control. *Chemometr. Intell. Lab.* **39,** 29–40.

Grung, B. and Manne, R. (1998). Missing Values in Principal Component Analysis. *Chemometr. Intell. Lab.* **42,** 125–139.

Gunderson, R.W., Thrane, K.R. and Nilson, R.D. (1988). A False-Colour Technique for Display and Analysis of Multivariate Chemometric Data. *Chemometr. Intell. Lab.* **3,** 119–131.

Halperin, M. (1970). On Inverse Estimators in Linear Regression. *Technometrics* **12,** 727–736.

Healy, M.J.R. (2000). *Matrices for Statistics*, 2$^{nd}$ Edn. Oxford University Press, Oxford, UK.

Helland, I.S. (1988). On the Structure of Partial Least Squares Regression. *Commun. Stat. Simulat.* **17,** 581–607.

Helland, I.S. and Almøy, T. (1994). Comparison of Prediction Methods when only a few Components are Relevant. *J. Am. Statist. Assoc.* **89,** 583–591.

Helland, I.S., Næs, T. and Isaksson, T. (1995). Related Versions of the Multiplicative Scatter Correction Methods for Pre-Processing Spectroscopic Data. *Chemometr. Intell. Lab.* **29,** 233–241.

Helland, K, Bentsen, H., Borgen, O.S. and Martens, H. (1992). Recursive Algorithm for Partial Least

Squares Regression. *Chemometr. Intell. Lab.* **14,** 129–137.

Herschel, W. (1800). Experiments on the Refrangibility of the Invisible Rays of the Sun. *Phil. Trans. Roy. Soc. London, Part II* 284–292.

Hertz, J., Krogh, A. and Palmer, R. (1991). *Introduction to the Theory of Neural Computation.* Addison-Wesley, Redwood City, CA, USA.

Hildrum, K.I., Martens, M. and Martens, H. (1983). Research on Analysis of Food Quality. *International Symp. on Control of Food and Food Quality*, Reading University, Reading, UK, March 22–24.

Honigs, D., Hieftje, G.M., Mark, H. and Hirschfeld, T.B. (1985). Unique Sample Selection via Near-Infrared Spectral Subtraction. *Anal. Chem.* **57,** 2299–2303.

Høskuldsson, A. (1988). PLS Regression Methods. *J. Chemometr.* **2,** 211–228.

Høskuldsson, A. (1996). *Prediction Methods in Science and Technology. Vol 1. Basic Theory.* Thor Publishing, Copenhagen, Denmark.

Indahl, U.G. and Næs, T. (1998). Evaluation of Alternative Spectral Feature Extraction Methods of Textural Images for Multivariate Modelling. *J. Chemometr.* **12,** 261–278.

Indahl, U.G., Sing, N.S., Kirkhuus, B. and Næs, T. (1999). Multivariate Strategies for Classification Based on NIR Spectra—with Applications to Mayonnaise. *Chemometr. Intell. Lab.* **49,** 19–31.

Isaksson, T. and Kowalski, B. (1993). Piece-Wise Multiplicative Scatter Correction Applied to Near-Infrared Diffuse Transmittance Data from Meat Products. *Appl. Spectrosc.* **47,** 702–709.

Isaksson, T. and Næs, T. (1988). The Effect of Multiplicative Scatter Correction and Linearity Improvement in NIR Spectroscopy. *Appl. Spectrosc.* **42,** 1273–1284.

Isaksson, T. and Næs, T. (1990). Selection of Samples for Calibration in Near-Infrared Spectroscopy. Part II: Selection based on Spectral Measurements. *Appl. Spectrosc.* **44,** 1152–1158.

Isaksson, T., Wang, Z. and Kowalski, B. (1993). Optimised Scaling (OS2) Regression Applied to Near Infrared Diffuse Spectroscopy Data from Food Products. *J. Near Infrared Spectrosc.* **1,** 85–97.

Jensen, S.Å., Munck, L. and Martens, H. (1982). The Botanical Constituents of Wheat and Wheat Milling Fractions. I. Quantification by Autofluorescence. *Cereal Chem.* **59,** 477–484.

Joliffe, I.T. (1986). *Principal Component Analysis.* Springer Verlag, New York, USA.

Karstang, T.V. and Manne, R. (1992). Optimised Scaling. A Novel Approach to Linear Calibration with Closed Data Sets. *Chemometr. Intell. Lab.* **14,** 165–173.

Kaufman, L. and Rousseeuw, P.J. (1990). *Finding Groups in Data.* John Wiley & Sons, New York, USA.

Kennard. R.W. and Stone, L.A. (1969). Computer Aided Design of Experiments. *Technometrics* **11,** 137–148.

Kubelka, P. and Munck, F. (1931). Ein Beitrag zur Optik Farbanstriche. *Z. technische Physik* **12,** 593–604.

Lambert, J. H. (1760). Photometria, Augsburg.

Lea, P., Næs, T. and Rødbotten, M. (1991). *Analysis of Variance for Sensory Data.* John Wiley & Sons, Chichester, UK.

Leardi, R., Boggia, R. and Terrile, M. (1992). Genetic Algorithms as a Strategy for Feature Selection. *J. Chemometr.* **6,** 267–282.

Leung, A.K.M., Chau, F.T. and Gao, J.B. (1998). A Review on Applications of Wavelet Transform Techniques in Chemical Analysis: 1989–1997. *Chemometr. Intell. Lab.* **43,** 165–184.

Lorber, A., Wangen, L. and Kowalski, B.R. (1987). A Theoretical Foundation for PLS. *J. Chemometr.* **1,** 19–31.

Mallows, C.L. (1973). Some Comments on $C_p$. *Technometrics* **15,** 661–675.

Manne, R. (1987). Analysis of Two Partial Least Squares Algorithms for Multivariate Calibration. *Chemometr. Intell. Lab.* **2,** 187–197.

Mardia, K.V., Kent, J.T. and Bibby, J.M. (1979). *Multivariate Analysis.* Academic Press, London, UK.

Mark, H. (1991). *Principles and Practice of Spectroscopic Calibration.* John Wiley & Sons, New York, USA.

Mark, H. and Workman, J (1988). A New Approach to Generating Transferable Calibrations for Quantitative Near-Infrared Spectroscopy. *Spectroscopy* **3(11)**, 28–36.

Martens, H., Jensen, S.Å. and Geladi, P. (1983). Multivariate Linearity Transformations for Near-Infrared Reflectance Spectrometry. In *Proceedings from Nordic Symposium on Applied Statistics, Stavanger, Norway*. Stokkand Forlag Publishers, Stavanger, Norway, pp. 205–233.

Martens, H. and Martens, M. (2000). Modified Jack-Knife of Parameter Uncertainty in Bilinear Modelling by Partial Least Squares Regression (PLSR). *Food Qual. Pref.* **11**, 5–16.

Martens, H. and Martens, M. (2001). *Multivariate Analysis of Quality*. John Wiley & Sons, Chichester, UK.

Martens, H. and Næs, T. (1987). Multivariate Calibration by Data Compression. In *Near Infrared Technology in the Agricultural and Food Industries*, Ed by Williams, P.C. and Norris, K. American Association of Cereal Chemists, St Paul, MN, USA, pp. 57–87.

Martens, H. and Næs, T. (1989). *Multivariate Calibration*. John Wiley & Sons, Chichester, UK.

Martens, H. and Stark, E.J. (1991). Extended Multiplicative Signal Correction and Spectral Interference Subtraction: New Preprocessing Methods for Near Infrared Spectroscopy. *Pharmaceut. Biomed. Anal.* **9**, 625–635.

McClure, W.F., Hamid, A., Giesbrecht, F.G. and Weeks, W.W. (1984). Fourier Analysis Enhances NIR Diffuse Reflectance Spectroscopy. *Appl. Spectrosc.* **38**, 322–329.

McCullagh, P. and Nelder, J.A. (1983). *Generalised Linear Models*. Chapman and Hall, London, UK.

McElhinney, J., Downey, G. and Fearn, T. (1999). Chemometric Processing of Visible and Near Infrared Reflectance Spectra for Species Identification in Selected Raw Homogenised Meats. *J. Near Infrared Spectrosc.* **7,** 145–154.

McLachlan, G.J. (1992). *Discriminant Analysis and Statistical Pattern Recognition*. John Wiley & Sons, Chichester, UK.

Miller, C.E. and Næs, T. (1990). A Path Length Correction Method for Near-Infrared Spectroscopy. *Appl. Spectrosc.* **44,** 895–898.

Miller, C.E., Svendsen, S. and Næs, T. (1993). Non-Destructive Characterisation of Polyethylene/Nylon Laminates by Near-Infrared Spectroscopy. *Appl. Spectrosc.* **47,** 346–356.

Miller, C.E. (1993). Sources of Non-Linearity in Near Infrared Methods. *NIR news* **4(6),** 3–5.

Montgomery, D.C. (1997). *Introduction to Statistical Quality Control*, 3$^{rd}$ Edn. John Wiley & Sons, New York, USA.

Næs, T. (1987). The Design of Calibration in Near Infrared Reflectance Analysis by Clustering. *J. Chemometr.* **1,** 121–134.

Næs, T. (1989). Leverage and Influence Measures for Principal Component Regression. *Chemometr. Intell. Lab.* **5,** 155–168.

Næs, T. (1991). Multivariate Calibration when Data are Split into Subsets. *J. Chemometr.* **5,** 487–501.

Næs, T. and Ellekjær, M.R (1992). The Relation between *SEP* and Confidence Intervals. *NIR news* **3(6)**, 6–7.

Næs, T. and Helland, I.S. (1993). Relevant components in regression. *Scand. J. Statist.* **20**, 239–250.

Næs, T. and Hildrum, K.I. (1997). Comparison of Multivariate Calibration and Discriminant Analysis in Evaluating NIR Spectroscopy for Determination of Meat Tenderness. *Appl. Spectrosc.* **51**, 350–357.

Næs, T. and Indahl, U. (1998). A Unified Description of Classical Methods for Multicollinear Data. *J. Chemometr.* **12**, 205–220.

Næs, T., Irgens, C. and Martens, H. (1986). Comparison of Linear Statistical Methods for Calibration of NIR Instruments. *Appl. Statist.* **35**, 195–206.

Næs, T. and Isaksson, T. (1989). Selection of Samples for Near-Infrared Spectroscopy, I. General Principle Illustrated by Examples. *Appl. Spectrosc.* **43**, 328–335.

Næs, T. and Isaksson, T. (1991). Splitting of Calibration Data by Cluster Analysis. *J. Chemometr.* **5**, 49–65.

Næs, T. and Isaksson, T. (1992a). Computer-Assisted Methods in Near Infrared Spectroscopy. In *Computer-Enhanced Analytical Spectroscopy, Vol. 3*, Ed by Jurs, P. Plenum Press, New York, USA, pp. 69–94.

Næs, T. and Isaksson, T. (1992b). Locally Weighted Regression in Diffuse Near-infrared Transmittance Spectroscopy. *Appl. Spectrosc.* **46**, 34–43.

Næs, T. Isaksson, T and Kowalski, B.R (1990). Locally Weighted Regression and Scatter-Correction for Near-Infrared Reflectance Data, *Anal. Chem.* **62**, 664–673.

Næs, T., Kvaal, K., Iskasson, T. and Miller, C. (1993). Artificial Neural Networks in Multivariate Calibration. *J. Near Infrared Spectrosc.* **1**, 1–11.

Næs, T. and Martens, H. (1988). Principal Component Regression in NIR Analysis. *J. Chemometr.* **2**, 155–167.

Nørgaard, L. (1995). Direct Standardization in Multiwavelength Fluorescence Spectroscopy. *Chemometr. Intell. Lab.* **29**, 283–293.

Nørgaard, L. and Bro, R. (1999). PLS Regression in the Food Industry. A Study of N-PLS Regression and Variable Selection for Improving Prediction Errors and Interpretation. In *Proceedings from Symposium on PLS methods, 1999*, Ed by Morineau, A. and Tenenhaus, M. Centre International de Statistique et d'Informatique Appliquees, 261 rue de Paris, 93556 Montreuil Cedex, France.

Ogden, R.T. (1997). *Essential Wavelets for Statistical Applications and Data Analysis*. Birkhauser, Boston, USA.

Oman, S. (1985). An Exact Formula for Mean Square Error of the Inverse Estimator in the Linear Calibration Problem. *J. Statist. Plan. Infer.* **11**, 189–196.

Oman, S., Næs, T. and Zube, A. (1993). Detecting and Adjusting for Nonlinearities in Calibration of Near-Infrared Data Using Principal Components, *J. Chemometr.* **7**, 195–212.

Osborne, B.G, Fearn, T. and Hindle, P.H. (1993). *Practical NIR Spectroscopy with Applications in Food and Beverage Analysis.* Longman Scientific & Technical, Harlow, UK.

Pao, Y.H. (1989). *Adaptive Pattern Recognition and Neural Networks.* Addison-Wesley, Reading, MA, USA.

Ripley, B.D. (1996). *Pattern Recognition and Neural Networks.* Cambridge University Press, Cambridge, UK.

Rousseeuw, P.J., Derde, M.-P. and Kaufman, L. (1989). Principal Components of a Fuzzy Clustering. *Trends Anal. Chem.* **8(7),** 249–250.

Rousseeuw, P.J. and Leroy, A.M. (1987). *Robust Regression and Outlier Detection.* John Wiley & Sons, New York, USA.

Sahni, N.S., Isaksson, T. and Næs, T. (2001). The use of experimental design methodology and multivariate analysis to determine critical control points in a process. *Chemometr. Intell. Lab.* **56(2),** 105–121 (2001).

Savitzky, A. and Golay, M.J.E. (1964). Smoothing and Differentiation of Data by Simplified Least-Squares Procedures. *Anal. Chem.* **36,** 1627–1639.

Searle, S.R. (1971). *Linear Models.* John Wiley & Sons, New York, USA.

Seculic, S., Seasholtz, M.B., Wang, Z.S.E., Holt, B.R. and Kowalski, B. (1993). Non-Linear Multivariate Calibration Methods in Analytical Chemistry. *Anal. Chem.* **65,** 835A–845A.

Shenk, J.S., Westerhaus, M.O. and Templeton, W.C. (1985). Calibration Transfer Between Near-Infrared Spectrophotometers. *Crop Sci.* **25,** 159–161.

Shenk, J.S., Westerhaus, M.O. and Berzaghi, P. (1997). Investigation of a LOCAL Calibration Procedure for Near Infrared Instruments. *J. Near. Infrared Spectrosc.* **5,** 223–232.

Shenk, J.S., Berzaghi, P. and Westerhaus, M.O. (2000). LOCAL: A Unifying Theory and Concept for Near Infrared Analysis. In *Near Infrared Spectroscopy: Proceedings of the 9th International Conference*, Ed by Davies, A.M.C. and Giangiacomo, R. NIR Publications, Chichester, UK.

Shukla, G.K. (1972). On the Problem of Calibration. *Technometrics* **14,** 547–553.

Snedecor, G.W. and Cochran, W.G. (1967). *Statistical Methods.* Iowa State University Press, Ames, IA, USA.

Steinier, J., Termonia, Y. and Deltour, J. (1972). Comments on Smoothing and Differentiation of Data by Simplified Least Squares Procedure. *Anal. Chem.* **44,** 1906–1909.

Stone, M. (1974). Cross-Validatory Choice and Assessment of Statistical Prediction. *J. R. Statist. Soc. B* **39,** 111–133.

Stone, M. and Brooks, R.J. (1990). Continuum Regression. Cross-Validated Sequentially Constructed Prediction Embracing Ordinary Least Squares, Partial Least Squares and Principal Components Regression. *J. R. Statist. Soc. B* **52,** 237–269; Corrigendum (1992), **54,** 906–907.

Sundberg. R. (1985). When is the Inverse Estimator MSE Superior to the Standard Regression Estimator in Multivariate Calibration Situations? *Statist. Prob. Lett.* **3,** 75–79.

Sundberg, R. (1999). Multivariate Calibration—Direct and Indirect Regression Methodology (with discussion). *Scand. J. Statist.* **26,** 161–207.

Thyholt, K. and Isaksson, T. (1997). Differentiation of Frozen and Unfrozen Beef Using Near-Infrared Spectroscopy. *J. Sci. Food Agric.* **73,** 525–532.

Velleman, P.F. and Welch, R.E. (1981). Efficient Computing of Regression Diagnostics. *Am. Stat.* **35,** 513–522.

Walczak, B., Bouveresse, E. and Massart, D.L. (1997). Standardization of Near-Infrared Spectra in the Wavelet Domain. *Chemometr. Intell. Lab.* **36,** 41–51.

Walczak, B. and Massart, D.L. (1997). Wavelets—Something for Analytical Chemistry? *Trends Anal. Chem.* **16,** 451–463.

Wang, Z., Isaksson, T. and Kowalski, B.R. (1994). New Approach for Distance Measurement in Locally Weighted Regression. *Anal. Chem.* **66,** 249–260.

Wang, Y. and Kowalski, B.R. (1992). Calibration Transfer and Measurement Stability of Near-Infrared Spectrometers. *Appl. Spectrosc.* **46,** 764–771.

Wang, Y., Lysaght, M.J. and Kowalski, B.R. (1993). Improvement of Multivariate Calibration Through Instrument Standardization. *Anal. Chem.* **64,** 562–564.

Wang, Y., Veltkamp, D.J. and Kowalski, B.R. (1991). Multivariate Instrument Standardization. *Anal. Chem.* **63,** 2750–2758.

Wedel, M. and Steenkamp, J.E.M. (1989). A Fuzzy Cluster-Wise Regression Approach to Benefit Segmentation. *Int. J. Res. Mark.* **6,** 241–258.

Weisberg, S. (1985). *Applied Linear Regression.* John Wiley & Sons, New York, USA.

Westad, F. and Martens, H. (2000). Variable Selection in NIR Based on Significance Testing in Partial Least Squares Regression (PLSR). *J. Near Infrared Spectrosc.* **8,** 117–124.

Williams, P.C. and Norris, K.H. (Eds) (1987). *Near Infrared Technology in the Agricultural and Food Industries.* American Association of Cereal Chemists, St Paul, MN, USA.

Windham, M.P. (1987). Cluster Validity for Fuzzy Clustering Algorithms. *Fuzzy Set. Syst.* **5,** 177–185.

Wold, S. (1976). Pattern Recognition by Means of Disjoint Principal Components Models. *Pattern Recogn.* **8,** 127–139.

Wold, S. (1978). Cross-Validatory Estimation of the Number of Components in Factor Analysis and Principal Components Models. *Technometrics* **20,** 397–406.

Wold. S. (1992). Nonlinear Partial Least Squares Modeling. II, Spline Inner Relation. *Chemometr. Intell. Lab.* **14,** 71–84.

Wold, S., Antii, S.H., Lindgren, F. and Öhman, J. (1998). Orthogonal Signal Correction of Near-

Infrared Spectra. *Chemometr. Intell. Lab.* **44,** 175–185.

Wold, S., Martens, H. and Wold, H. (1984). The Multivariate Calibration Problem in Chemistry Solved by the PLS Method. In *Proc. Conf. Matrix Pencils*, Ed by Ruhe, A. and Kagstrom, B. Lecture Notes in Mathematics. Springer Verlag, Heidelberg, Germany, pp. 286–293.

Wold, S. and Sjøstrøm, M. (1977). SIMCA. A Method for Analyzing Chemical Data in Terms of Similarity and Analogy. In *Chemometrics: Theory and Publications*, Ed by Kowalski, B.R. American Chemical Society, Washington, USA, pp. 243–282.

Zemroch, P.J. (1986). Cluster Analysis as an Experimental Design Generator, with Application to Gasoline Blending Experiments. *Technometrics* **28,** 39–49.

# A Appendix A. Technical details

## A.1 An introduction to vectors and matrices

What follows is a brief introduction to vectors and matrices, explaining some of the notation used in the book and describing the basic operations of matrix algebra. For a much fuller treatment see, for example, Healy (2000).

A single number, e.g. 5, 12.4, $3 \times 10^5$, is called a scalar. If we want to refer to such a quantity without specifying its value we use a lower case italic letter, e.g. let $x$ be the pH of the sample.

A vector, denoted here by a lower case bold letter, is an array of scalars written as a column

$$\mathbf{x} = \begin{bmatrix} 2 \\ 1 \\ 4 \\ 3 \end{bmatrix} \quad (A1)$$

**SVD**: singular value decomposition

It is also possible to write a vector as a row, but in this book, unless otherwise indicated, a vector will be a column.

A general column vector $\mathbf{x}$ with $N$ elements is usually written

$$\mathbf{x} = \begin{bmatrix} x_1 \\ \cdot \\ x_i \\ \cdot \\ x_N \end{bmatrix} \quad (A2)$$

with $x_i$ denoting the $i$th element. For example $\mathbf{x}$ might be the spectrum of a sample, with $x_i$ the absorbance at the $i$th of $N$ wavelengths.

The transpose operation, denoted by a superscript $t$, converts a column into a row, so for the vector in (A1)

$$\mathbf{x}^t = [2 \quad 1 \quad 4 \quad 3] \tag{A3}$$

A vector with $N$ elements can be represented geometrically in an $N$-dimensional space, either as the point in this space whose coordinates are given by the elements, or as the line joining the origin of the axes to this point. For $N = 3$, this can be visualised as in Figure A1. For $N > 3$, this visualisation is not possible, but the mathematics is the same and the interpretations carry over.

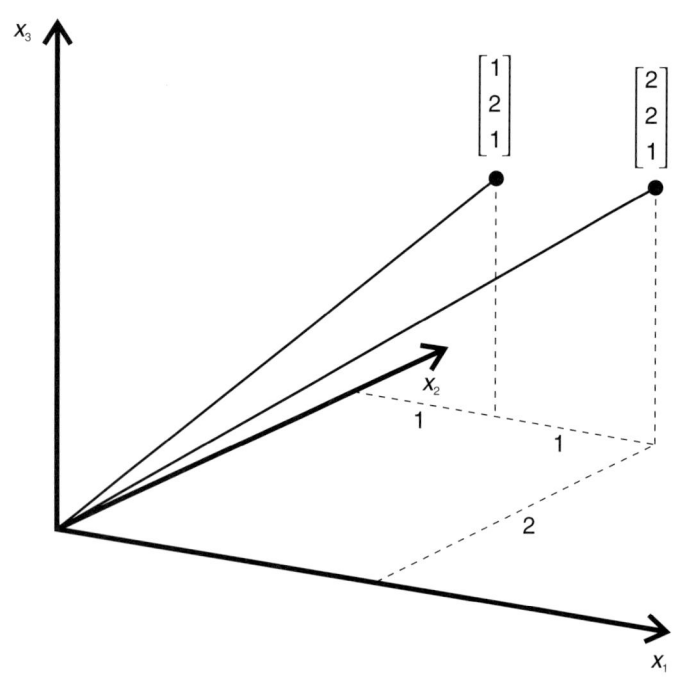

**Figure A1.** Geometrical representation of two 3-dimensional vectors [2 2 1]$^t$ and [1 2 1]$^t$. The three arrows represent the axes in the 3-dimensional coordinate system.

So far as the representation is concerned, it makes no difference whether a vector is written as a row or a column. When it comes to doing matrix algebra, however, it does matter. As we shall see, the shapes of matrices (and vectors) have to match for them to be combined by addition or multiplication. When **x** is just described as a vector the usual assumption is that it is a column vector.

Vectors are added element-wise, so, for example,

$$\begin{bmatrix} 2 \\ 2 \\ 1 \end{bmatrix} + \begin{bmatrix} 1 \\ 2 \\ 1 \end{bmatrix} = \begin{bmatrix} 3 \\ 4 \\ 2 \end{bmatrix} \quad (A4)$$

This has a simple geometric interpretation as addition of the lines in Figure A1. If we take either one of the two lines shown and shift it (keeping its orientation the same) so that it starts where the other ends, then its far end will be the point representing the sum of the vectors. Do it both ways and the result is a parallelogram whose diagonal is the sum of the two vectors. We can also multiply a vector by a scalar, an operation that is again performed element-wise, so that

$$0.5 \times \begin{bmatrix} 2 \\ 2 \\ 1 \end{bmatrix} = \begin{bmatrix} 1 \\ 1 \\ 0.5 \end{bmatrix} \quad (A5)$$

is a vector with the same direction as the original but half the length. Multiplying by a negative number reverses the direction, and changes the length as well unless the number is $-1$.

A pair of vectors **x** and **y** in a three-dimensional space, such as those in Figure A1, define a plane, in the sense that there is a unique plane in which both of them lie. They are also described as spanning the subspace of all vectors that lie in the plane, meaning

**A.1** An introduction to vectors and matrices

that any vector in the plane can be represented as a linear combination $a\mathbf{x} + b\mathbf{y}$ where $a$ and $b$ are scalars. This idea generalises to higher dimensions. A set of $p$ vectors in an $N > p$ dimensional space will span a subspace, described as a hyperplane. The dimension of this subspace may be as great as $p$. It will be less if some of the vectors can be expressed as linear combinations of the others. If we have as many vectors as dimensions of the space, three vectors in three-dimensional space for example, then they will span the whole space unless one can be represented as a linear combination of the other two, i.e. one lies in the plane defined by the other two, in which case they only span a two-dimensional plane, or all three have exactly the same direction, in which case they span only a one-dimensional subspace.

A matrix is an $N \times K$ array of scalars, and will be denoted here by a bold capital letter. We denote the general element by $x_{ik}$, using two subscripts to indicate row and column:

$$\mathbf{X} = \begin{bmatrix} x_{11} & . & x_{1K} \\ . & x_{ik} & . \\ x_{N1} & . & x_{NK} \end{bmatrix} \qquad (A6)$$

It is possible to regard such a matrix as just an array of scalars, or as a set of $N$ row vectors, each of length $K$, or as a set of $K$ column vectors, each of length $N$. All of these interpretations can be useful, depending on the context. For example if $x_{ik}$ is an absorbance measurement at wavelength $k$ on sample $i$ then the rows are the spectra of individual samples whilst the columns are measurements at one wavelength for all samples.

Regarding the matrix as a collection of row or column vectors, it is sometimes useful to consider the subspace spanned by these vectors. For example, if the rows of $\mathbf{X}$ are spectra, and the samples are mixtures of a small number of components, and the system is linear and additive, the measured spectra will be linear combinations of the spectra of the "pure" components (Beer's law), and the subspace spanned by the rows will be of very low dimension. A general (and non-obvious) result is that the subspaces spanned by the rows and columns of a matrix have the same dimension.

We will now describe some basic mathematical operations on matrices that are used frequently throughout the book: transpose, addition and subtraction, multiplication and inversion.

We have already seen how to transpose vectors. The effect on a matrix is the same, with the rows being converted into columns (and vice versa). For a $(2 \times 3)$ matrix $\mathbf{X}$ this can be visualised as:

$$\mathbf{X} = \begin{bmatrix} x_{11} & x_{12} & x_{13} \\ x_{21} & x_{22} & x_{23} \end{bmatrix}_{(2 \times 3)} \Leftrightarrow \mathbf{X}^t = \begin{bmatrix} x_{11} & x_{21} \\ x_{12} & x_{22} \\ x_{13} & x_{23} \end{bmatrix}_{(3 \times 2)} \quad (A7)$$

Addition (or subtraction) of two equal sized matrices, here $(3 \times 2)$ matrices, is performed element-wise, as with vectors, so that:

$$\mathbf{X} = \begin{bmatrix} x_{11} & x_{12} \\ x_{21} & x_{22} \\ x_{31} & x_{32} \end{bmatrix} \text{ and } \mathbf{Y} = \begin{bmatrix} y_{11} & y_{12} \\ y_{21} & y_{22} \\ y_{31} & y_{32} \end{bmatrix}$$

$$\Rightarrow \mathbf{X} + \mathbf{Y} = \begin{bmatrix} x_{11} + y_{11} & x_{12} + y_{12} \\ x_{21} + y_{21} & x_{22} + y_{22} \\ x_{31} + y_{31} & x_{32} + y_{32} \end{bmatrix} \quad (A8)$$

Subtraction is done the same way.

Multiplication of two matrices, here one $(2 \times 3)$ and one $(3 \times 2)$ matrix, is not done element-wise. It is defined as:

**A.1 An introduction to vectors and matrices**

$$\mathbf{X} = \begin{bmatrix} x_{11} & x_{12} & x_{13} \\ x_{21} & x_{22} & x_{23} \end{bmatrix}_{(2 \times 3)} \text{ and } \mathbf{Y} = \begin{bmatrix} y_{11} & y_{12} \\ y_{21} & y_{22} \\ y_{31} & y_{32} \end{bmatrix}_{(3 \times 2)} \Rightarrow$$

$$\mathbf{X} * \mathbf{Y} = \begin{bmatrix} x_{11} \times y_{11} + x_{12} \times y_{21} + x_{13} \times y_{31} & x_{11} \times y_{12} + x_{12} \times y_{22} + x_{13} \times y_{32} \\ x_{21} \times y_{11} + x_{22} \times y_{21} + x_{23} \times y_{31} & x_{21} \times y_{12} + x_{22} \times y_{22} + x_{23} \times y_{32} \end{bmatrix}_{(2 \times 2)} \quad (A9)$$

The number of columns in the first matrix (here 3) must correspond to the number of rows in the second leading to a product with dimensions corresponding to the number of rows in the first (here 2) and the number of columns in the second (here 2). We have used a * to denote the matrix product above, but it is more usual to employ no symbol at all and write the product simply as **XY**.

The one exception to this rule that dimensions must match is that we may multiply a matrix **X** of any size by a scalar $y$. The result is defined to be the matrix obtained by multiplying each element of **X** by $y$.

Vectors are just matrices with one dimension equal to 1, and the same multiplication rule applies. It is interesting to look at the effect of multiplying a vector by a matrix. If **X** is a $p \times q$ matrix and **v** is a $q \times 1$ vector, the product

$$\mathbf{w} = \mathbf{Xv} \quad (A10)$$

is a $p \times 1$ vector, each of whose elements is a linear combination of the elements of **v**. A simple example, with **X** a $3 \times 3$ matrix and **v** and **w** $3 \times 1$ vectors is shown below.

$$\begin{bmatrix} 0.707 & -0.707 & 0 \\ 0.707 & 0.707 & 0 \\ 0 & 0 & -1 \end{bmatrix} \begin{bmatrix} 1 \\ 1 \\ 1 \end{bmatrix} = \begin{bmatrix} 0 \\ 1.414 \\ -1 \end{bmatrix} \quad (A11)$$

Considering their action on vectors gives us another way of thinking of matrices. We can view them as linear transformations that transform, or "map", vectors into other vectors. If **X** is square, as it is in the example above, both vectors lie in the same dimensional space. If it is not square then the result is either a shorter or a longer vector. One special square matrix is the identity matrix, usually denoted by **I**, whose elements are 1 on the diagonal and 0 everywhere else. It is easy to see that the identity matrix maps any vector into itself, or

$$\mathbf{Iv} = \mathbf{v} \tag{A12}$$

Thinking of matrices as linear transformations, we can interpret the product $\mathbf{Z} = \mathbf{XY}$ of two matrices **X** and **Y** as the linear transformation that has the same effect as applying first **Y** and then **X**, so that if $\mathbf{Yu} = \mathbf{v}$ and $\mathbf{Xv} = \mathbf{w}$ then $\mathbf{Zu} = \mathbf{XYu} = \mathbf{Xv} = \mathbf{w}$. In words, we can either apply **Y** then **X** to **u** or we can multiply **X** and **Y** using matrix multiplication to get **Z** and then apply **Z** to **u**, the effect will be the same.

Still thinking of matrices as linear transformations, and particularly of square matrices because these preserve the dimension of the vectors they act on, it is natural to ask whether, for a given matrix **X**, there exists a transformation that reverses the effect of **X**. Is there a matrix **Y** such that

$$\mathbf{YXv} = \mathbf{v} \tag{A13}$$

for all vectors **v**? If **X** is square and such a matrix **Y** exists then it is unique, it is called the inverse of **X**, and it is denoted by $\mathbf{X}^{-1}$. Since $\mathbf{X}^{-1}\mathbf{Xv} = \mathbf{v}$ for all vectors **v**, we must have $\mathbf{X}^{-1}\mathbf{X} = \mathbf{I}$, and it is also easy to show that $\mathbf{XX}^{-1} = \mathbf{I}$, i.e. that **X** is the inverse of $\mathbf{X}^{-1}$.

The inverse of the matrix in Equation (A11) is

$$\begin{bmatrix} 0.707 & 0.707 & 0 \\ -0.707 & 0.707 & 0 \\ 0 & 0 & -1 \end{bmatrix} \tag{A14}$$

## A.1 An introduction to vectors and matrices

as can be checked by multiplying the two matrices together.

Not all square matrices have inverses. If $\mathbf{X}$ maps two different vectors $\mathbf{u}$ and $\mathbf{v}$ to the same result $\mathbf{w}$, so that $\mathbf{Xu} = \mathbf{Xv} = \mathbf{w}$, then it cannot have an inverse, for we would need $\mathbf{X}^{-1}\mathbf{w} = \mathbf{u}$ and $\mathbf{X}^{-1}\mathbf{w} = \mathbf{v}$ at the same time. Such a matrix, $\mathbf{X}$, is called singular. It maps the non-zero vector $\mathbf{y} = \mathbf{u} - \mathbf{v}$ to the zero vector $\mathbf{0} = [0, ..., 0]'$, i.e. $\mathbf{Xy} = \mathbf{0}$ for some non-zero $\mathbf{y}$. Since the vector $\mathbf{Xy}$ is a linear combination of the columns of $\mathbf{X}$, $\mathbf{Xy} = \mathbf{0}$ implies linear dependencies amongst these columns. Thus a square matrix is singular if the subspace spanned by its columns (or rows) is less than the full space. If multiplying by $\mathbf{X}$ compresses all vectors into a lower dimensional space than the full one, we cannot recover all the original vectors, and therefore $\mathbf{X}$ has no inverse. The dimension of the subspace spanned by the rows (or columns) of a matrix is called its rank. A square matrix with rank = dimension, and which therefore has an inverse, is called "full rank".

One important application of matrix inverses is to the solution of sets of simultaneous linear equations. We can write a set of $p$ linear equations in $p$ unknowns as

$$\mathbf{Ay} = \mathbf{b} \qquad (A15)$$

where $\mathbf{A}$ is a $p \times p$ matrix of coefficients, one row of $\mathbf{A}$ containing the coefficients of one equation, $\mathbf{y}$ is a $p \times 1$ vector of unknowns, and $\mathbf{b}$ is a $p \times 1$ vector of constants, the right hand sides of the equations. Then the solution, if $\mathbf{A}^{-1}$ exists, is

$$\mathbf{y} = \mathbf{A}^{-1}\mathbf{b} \qquad (A16)$$

Two vectors of the same length can be multiplied together in two different ways. If $\mathbf{x}$ and $\mathbf{y}$ are both $p \times 1$ vectors, the outer product $\mathbf{xy}'$ is a $p \times p$ matrix with $ik$th element $x_i y_k$, whilst the inner product $\mathbf{x}'\mathbf{y} = x_1 y_1 + ... + x_p y_p$ is a scalar. Two special cases are of interest. The product $\mathbf{x}'\mathbf{x} = x_1^2 + \ldots + x_p^2$ is the squared length of the vector $\mathbf{x}$. For example the two vectors in Figure A1 have lengths $\sqrt{6}$ and 3. If the length of a vector is equal to 1, it is sometimes called a unit length vector or just a unit vector. When $\mathbf{x}'\mathbf{y} = 0$ the vectors $\mathbf{x}$ and $\mathbf{y}$ are said to be orthogonal. Geometrically this means that the angle between them is a right angle.

Some square matrices are also described as orthogonal. For such a matrix, each of its rows, thought of as a vector, is orthogonal to all of the others and has length 1. These conditions imply that for an orthogonal matrix $\mathbf{P}$ we have

$$\mathbf{PP}' = \mathbf{I} \qquad (A17)$$

Since inverses are unique, we may deduce that $\mathbf{P}^{-1} = \mathbf{P}'$ and hence that

$$\mathbf{P}'\mathbf{P} = \mathbf{I} \qquad (A18)$$

also, and thus that the columns of $\mathbf{P}$, as well as the rows, must be mutually orthogonal and of length 1.

Orthogonal matrices have a simple geometrical interpretation as transformations. They are rigid rotations. If the two vectors in Figure A1 were subjected to a transformation represented by an orthogonal matrix their lengths and the angle between them would stay the same, but they would be rotated into another direction. The matrix in Equation (A11) is orthogonal (or would be if the elements were given exactly, 0.707 is an approximation to $1/\sqrt{2} = 0.7071...$), and its inverse, given in (A14), is also its transpose. It rotates vectors through 45° with respect to the first two axes and 90° with respect to the third.

A special type of linear transformation much used in chemometrics is the projection operator $\mathbf{P}_\mathbf{X}$. This transformation is defined relative to a particular $N \times K$ matrix $\mathbf{X}$ as

**A.1 An introduction to vectors and matrices**

$$\mathbf{P}_X = \mathbf{X}(\mathbf{X}'\mathbf{X})^{-1}\mathbf{X}' \qquad (A19)$$

and can be interpreted geometrically as the transformation that projects vectors onto the subspace spanned by the columns of $\mathbf{X}$. It is assumed here that $N > K$, so that the columns of $\mathbf{X}$ cannot span the whole space. The projection of a vector onto a subspace is the vector ending at the nearest point in the subspace to the end of the vector being projected. This is shown geometrically for an example in Figure A2.

Vectors already in the subspace are mapped into themselves. One consequence of this is that if $\mathbf{P}_X$ is applied a second time, nothing further happens, i.e. $\mathbf{P}_X \mathbf{P}_X = \mathbf{P}_X$.

Let $\mathbf{A}$ be a square matrix, then a vector $\mathbf{v}$ is called an eigenvector of $\mathbf{A}$ if

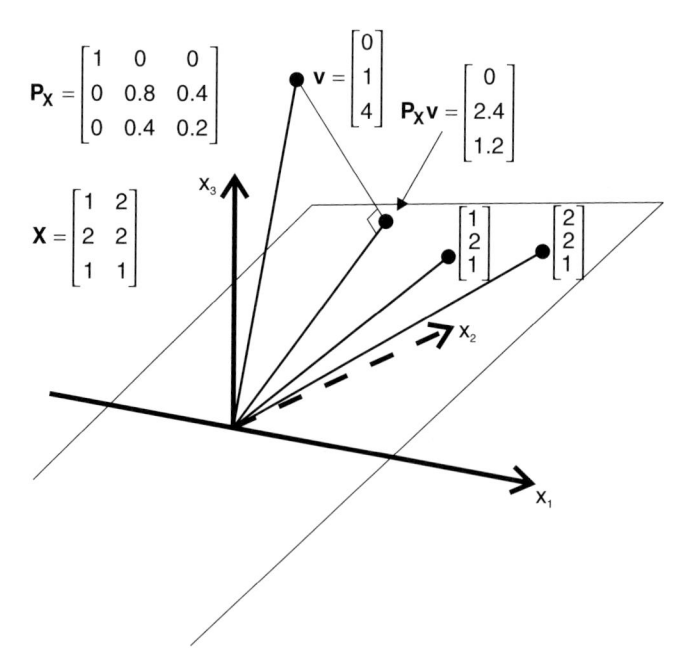

$$\mathbf{P}_X = \begin{bmatrix} 1 & 0 & 0 \\ 0 & 0.8 & 0.4 \\ 0 & 0.4 & 0.2 \end{bmatrix} \quad \mathbf{v} = \begin{bmatrix} 0 \\ 1 \\ 4 \end{bmatrix} \quad \mathbf{P}_X \mathbf{v} = \begin{bmatrix} 0 \\ 2.4 \\ 1.2 \end{bmatrix}$$

$$\mathbf{X} = \begin{bmatrix} 1 & 2 \\ 2 & 2 \\ 1 & 1 \end{bmatrix}$$

**Figure A2.** Geometrical illustration of subspace (plane) spanned by the vectors $[1,2,1]^t$ and $[2,2,1]^t$ and a projection of $\mathbf{v} = [0,1,4]^t$ onto the space. Again the three arrows represent the three coordinate axes.

$$\mathbf{A}\mathbf{v} = \lambda \mathbf{v} \quad (A20)$$

The $\lambda$ is called the corresponding eigenvalue. If the $\mathbf{A}$ is a covariance matrix (see Appendix A2) of a vector variable, the eigenvectors and eigenvalues have an interesting interpretation in terms of directions of main variability and their corresponding variance. This will be discussed next [see also Appendix A5 and Mardia *et al.* (1979)].

It can be shown that any matrix $\mathbf{X}$ can be decomposed as

$$\mathbf{X} = \hat{\mathbf{U}}\hat{\mathbf{S}}\hat{\mathbf{P}}^t \quad (A21)$$

Here $\hat{\mathbf{S}}$ is a matrix of zeros except for the diagonal containing the so-called singular values computed as square roots of the eigenvalues of $\mathbf{X}^t\mathbf{X}$, $\hat{\mathbf{U}}$ is an orthogonal matrix whose columns are the eigenvectors of $\mathbf{X}\mathbf{X}^t$ and $\hat{\mathbf{P}}$ is another orthogonal matrix whose columns are the eigenvectors of the matrix $\mathbf{X}^t\mathbf{X}$. Note that for a mean-centred $\mathbf{X}$-matrix, the $\mathbf{X}^t\mathbf{X}$ matrix is equal to $(N-1)$ times the covariance matrix (see A26). The decomposition (A21) is called the singular value decomposition (SVD) of $\mathbf{X}$.

The singular values and eigenvectors play a very important role in this book. In most cases, the SVD is truncated by leaving out components corresponding to the smallest diagonal values in $\hat{\mathbf{S}}$, ending up with models like the one in equation (5.3). If the matrix $\mathbf{X}$ is centred, the eigenvectors $\hat{\mathbf{P}}$ used in the truncated version define the most important directions of variability (see, for example, Chapter 5.1). The singular values correspond to the variability (variance) along these axes. In Chapter 5 the scores matrix $\hat{\mathbf{T}}$ is the product of the matrices $\hat{\mathbf{U}}$ and $\hat{\mathbf{S}}$ in the SVD. Often the decomposition in equation (5.3) is also called the principal component decomposition. We will not go into detail about the mathematical and computational aspects here, but instead refer the reader to Chapter 5.1

**A.1** An introduction to vectors and matrices

and Appendix A5 for an intuitive description of the concepts and their importance in calibration. For a more extensive treatment of SVD and the closely related principal component analysis (PCA), we refer to Mardia *et al.* (1979).

The sum of the eigenvalues of $\mathbf{X}'\mathbf{X}$ is equal to the sum of its diagonal elements. This diagonal sum can be interpreted as the total sum of squares of all the original variables. This result makes it natural to use the term percent explained variance for the ratio of the sum of eigenvalues for the components used in the PCA decomposition divided by the total sum of the eigenvalues. If for instance the ratio between the sum of the two largest eigenvalues and the total sum is equal to 90%, we say that the two first components describe 90% of the total variability in the data.

Methods for computing PCA in situations with missing data can be found in Grung and Manne (1998).

## A.2 Covariance matrices

Measures of location and spread of a variable $x$ are probably the most widely used quantities in statistics. In this book they play a crucial role. This section is devoted to a short discussion of some of the most used measures.

The location of a variable $x$ is most often measured by its mean or average value

$$\bar{x} = \sum_{i=1}^{N} x_i / N,$$

where $N$ is the number of measurements. Depending on the context, this may be though of either just as a convenient summary statistic, or as an estimate of some underlying population mean, the observed $x_i$ being imagined to be a sample from some infinite population of possible measurements.

# Appendix A. Technical details

The spread of a variable is most often measured by its variance or standard deviation. The variance of a variable $x$ is defined as

$$\text{var}(x) = \sum_{i=1}^{N}(x_i - \bar{x})^2 /(N-1) \quad (A22)$$

The standard deviation, std, is defined as the square root of the variance. As for the mean, the variance and standard deviation can be considered as empirical estimates of population analogues. The advantage of the standard deviation is that it is in the same measurement units as the original $x$-measurements. The variance is often denoted by $s^2$ or $\hat{\sigma}^2$. In some cases, an $x$ is attached, for instance $s_x^2$ to indicate the variance corresponds to variable $x$.

In multivariate analysis, one measures several variables, $x_1, \ldots, x_K$ for each of a number of objects $N$. Each of these variables has a mean and a variance (and hence a standard deviation). In addition, a covariance between each pair of variables can be defined. Covariances are measures of degree of co-variation among the variables and are defined as

$$\text{cov}(x_p, x_q) = \sum_{i=1}^{N}(x_{ip} - \bar{x}_p)(x_{iq} - \bar{x}_q)/(N-1) \quad (A23)$$

where $x_{ip}$ is the value of variable $p$ for object $i$, and $\bar{x}_p$ is the mean of variable $p$. Correlations are defined by dividing the covariances by the product of the standard deviations of the variables. Correlation is also a measure of degree of co-variation. More precisely, it is a measure of the degree of linear relationship between variables. The empirical correlation is usually denoted by r and is defined as

$$r(x_p, x_q) = \text{cov}(x_p, x_q)/[\text{std}(x_p)\text{std}(x_q)] \quad (A24)$$

Correlation has the advantage over covariance that it is independent of the units used for the vari-

## A.2 Covariance matrices

ables. The correlation takes on values in the interval [−1,1]. A correlation close to one indicates a positive relationship which is close to linear. A positive relationship here means that large values of one of the variables correspond to large values of the other and vice versa. A correlation close to −1 indicates a strong negative relationship. A correlation close to 0 indicates a very weak linear relationship between the variables.

Covariances and variances of a set of variables $x_1, \ldots, x_K$ are sometimes put into a matrix. This is called the covariance matrix of the vector of variables and is defined as

$$\hat{\Sigma} = \begin{pmatrix} \mathrm{var}(x_1) & \mathrm{cov}(x_1,x_2) & . & \mathrm{cov}(x_1,x_K) \\ \mathrm{cov}(x_2,x_1) & \mathrm{var}(x_2) & . & . \\ . & . & . & . \\ \mathrm{cov}(x_K,x_1) & . & . & \mathrm{var}(x_K) \end{pmatrix} \quad (A25)$$

It is easily seen from the definition in (A23) that this matrix is symmetric. It is used in many places in this book. In Chapter 18, for example, covariance matrices are used extensively.

If we let **X** be the matrix of mean centred **x**-values (i.e. each variable is corrected for its average, so that it has average equal to 0), the covariance matrix can also be written as

$$\hat{\Sigma} = \mathbf{X}'\mathbf{X}/(N-1) \quad (A26)$$

## A.3  Simple linear regression

In this section we will discuss some very basic ideas and methods related to simple linear regression. Simple linear regression is the simplest of all calibration/regression methods.

See Draper and Smith (1981) and Weisberg (1985) for a more thorough treatment of this methodology.

### A.3.1 An example: boiling point and pressure

We will illustrate the use of regression by using some classical literature data.

The data in Table A1 are the results of 17 experiments carried out by the Scottish physicist James Forbes. He measured the boiling point of water, using an apparatus of his own devising, and atmospheric pressure, using a barometer, at various locations in the Alps and Scotland. His aim was to be able to estimate pressure (and hence altitude) by boiling water, wristwatch-sized barometers not being available to nineteenth century alpinists. The data can be found in Forbes (1857), or perhaps more easily in Brown (1993).

In Figure A3 are plotted the pressures, on the $y$-axis, against the boiling points, on the $x$-axis. There is a fairly convincing linear relationship between the two measurements, all but one of which lie quite close to the straight line that has been drawn through them.

### A.3.2 Fitting a straight line by least squares

If the equation of the straight line is written

$$y = b_0 + b_1 x, \qquad (A27)$$

the observation $y$ is often described by the model

$$y = b_0 + b_1 x + f \qquad (A28)$$

Here the constants $b_0$ and $b_1$ are the parameters of the line, its intercept and slope respectively and $f$ is a random error term. Usually the $x$-variable is called the independent or the explanatory variable and $y$ is called the dependent variable.

**A.3** Simple linear regression

**Table A1. Pressures (in inches of mercury) and boiling points of water (in degrees Fahrenheit) in 17 locations.**

| Experiment | Pressure (in Hg) | Boiling point (°F) |
|---|---|---|
| 1 | 20.79 | 194.5 |
| 2 | 20.79 | 194.3 |
| 3 | 22.40 | 197.9 |
| 4 | 22.67 | 198.4 |
| 5 | 23.15 | 199.4 |
| 6 | 23.35 | 199.9 |
| 7 | 23.89 | 200.9 |
| 8 | 23.99 | 201.1 |
| 9 | 24.02 | 201.4 |
| 10 | 24.10 | 201.3 |
| 11 | 25.14 | 203.6 |
| 12 | 26.57 | 204.6 |
| 13 | 28.49 | 209.5 |
| 14 | 27.76 | 208.6 |
| 15 | 29.04 | 210.7 |
| 16 | 29.88 | 211.9 |
| 17 | 30.06 | 212.2 |

For given values of $b_0$ and $b_1$, i.e. for a given choice of line, we can plug the boiling point, $x_i$, for the $i$th item into the equation to get a predicted value

$$\hat{y}_i = b_0 + b_1 x_i \qquad (A29)$$

for pressure and a residual

# Appendix A. Technical details

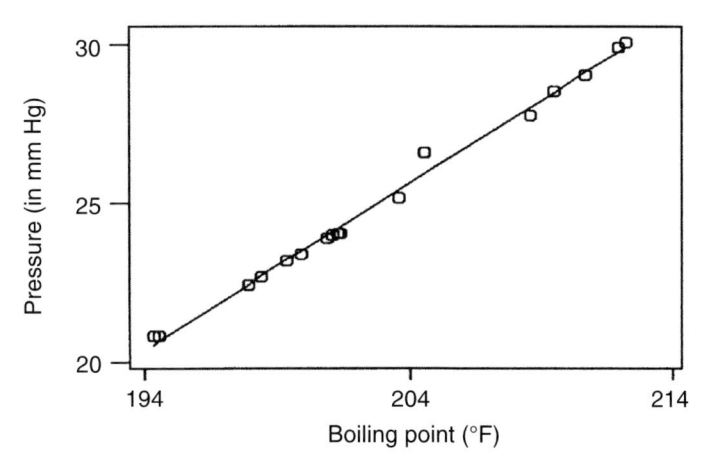

**Figure A3.** Forbes's data, plot of boiling point versus pressure.

$$\hat{f}_i = y_i - \hat{y}_i \tag{A30}$$

The residual is the difference between the measured pressure and that predicted by the equation. The sum over all $N = 17$ observations of the squared residuals,

$$RSS = \sum_{i=1}^{N} \hat{f}_i^2 \tag{A31}$$

is an overall measure of the goodness of fit of the line to the data. As we vary the coefficients, *RSS* will also vary. The principle of least squares (LS) chooses the values of $b_0$ and $b_1$ that give the smallest *RSS*. It ends up with estimates of the parameters denoted by $\hat{b}_0$ and $\hat{b}_1$. Later we will see how the least squares method can be used to fit more complicated equations. The basic principle is always the same: for fixed parameters; a fitted value and residual can be calculated for each sample, and we choose the parameter values that minimise the sum of the squared residuals (see Figure A4).

**A.3 Simple linear regression**

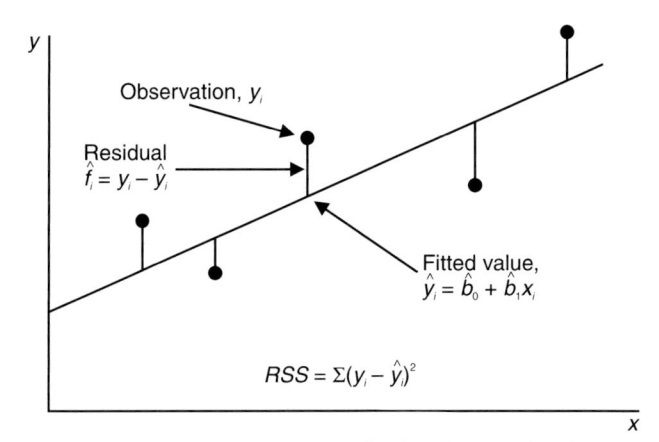

**Figure A4.** Fitted values and residuals. The LS criterion minimises the RSS (given in the illustration) over all possible parameter values.

### A.3.3 Models and errors

Why is least squares such a popular fitting method and when is it appropriate? The short answer to the first question is that it is popular because it is easy to implement. For a whole class of equations that are linear in their parameters it is possible to write down a closed form expression for the least squares estimates of the parameters and compute these estimates rapidly (see below).

The usual assumption for the error terms is that they are sampled from a normal or Gaussian population distribution with mean zero and the same amount of variability for all data points (same variance $\sigma^2$). If this assumption is reasonable then it can be shown that least squares is a good way to estimate the systematic part of the model. The most serious way in which the normality assumption can be violated is by error distributions that include occasional large errors—so-called outliers. When a residual $\hat{f}_i$ is large its square will be very large, and the least squares criterion *RSS* can easily be dominated by one or more large errors. Put another way, least squares is not robust to outliers, because it is tailored to normal data and normal data do not include outliers. Since most of the more sophisticated methods that will be met elsewhere in this book have a least squares fit hidden somewhere inside, this means that it will always be important to check for outliers.

There are other fitting methods, known as robust methods [see, for example, Rousseeuw and Leroy (1987)], that will cope with less well behaved error distributions. The difficulty with these is that they tend to be computationally expensive even for simple linear regression.

### A.3.4 Assessing the fit graphically

Having fitted a line we usually want to assess the quality of the fit. A simple graphical method of doing this is to plot the residuals against $x$, as in Figure A5. The point of doing this is to look for systematic patterns in the residuals. They ought to look random. Important departures are obvious curvature, outliers, i.e. points with large residuals, and non-constant scatter about the line. The residual plot for Forbes's data shows two of these problems.

The curvature in Figure A5 can, once you realise you are looking for it, be seen in Figure A3. The data

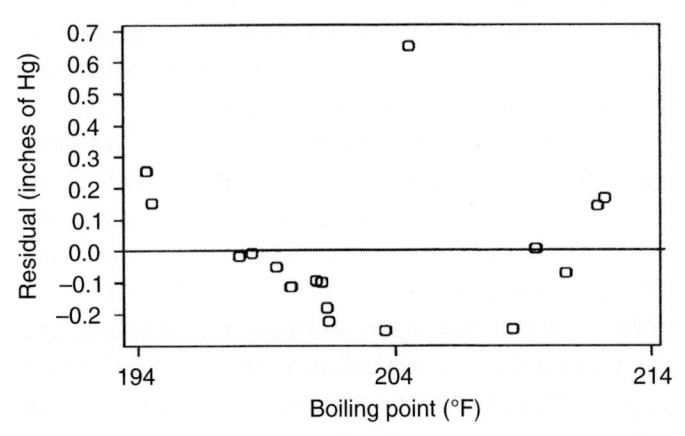

**Figure A5.** Forbes's data, plot of residuals versus boiling point.

**A.3** Simple linear regression

points at each end are above the fitted line; those in the middle are mostly below it. The residual plot rotates the line to the horizontal and blows up the vertical scale, making the pattern easier to see. Although the straight line fits the data quite well, we are missing an opportunity to do even better by taking more of the variability into the systematic part of our model and leaving less in the noise. One way to do this would be to fit a non-linear curve. A better way would be to look in a physics textbook and discover that there are good reasons for expecting the logarithm of pressure to have a linear relationship with the reciprocal of absolute temperature, as pointed out by Brown (1993) in his analysis of these data.

The data point at the top of Figure A5, the one with the largest residual, is clearly an outlier. Were the data not 150 years old, one might go back to the laboratory notebooks and check up on this point, or perhaps repeat the observation. Failing this, deleting the point from the data set seems the only real option. When models are fitted by least squares it is important not to have serious outliers present as they may grossly distort the fit. However, what counts as an outlier is not always easy to decide. The statistical literature is full of alternative criteria, and regression software will often flag points as outliers on the basis of one or more of these. Whether to remove them is a matter of judgement and unfortunately the advice to "just remove the obvious ones" is easier to give than to implement. Wholesale deletion of data points will certainly lead to over optimistic conclusions about how well the equation fits and should be avoided. However, even the single outlier that obviously needs to be removed may have an important message: that the equation does not work in every situation, for example with samples of a particular type or from a particular source.

A different, though related, idea is that of leverage. A point with high leverage is one whose $x$ value is far away from the mean $x$. Such points are very informative and therefore may have a big influence on the fitted line. If you have any doubts about the quality of the data for such a point then you need to worry. The outlier in Figure A5 has a boiling point close to the mean and thus very little leverage. Deleting it changes the intercept of the fitted line a little, and the slope hardly at all. Were one of the points at the end of the line to be displaced vertically by as much as this outlier it would noticeably change the slope of the least squares line. What is more, because the line "follows" the shift in the high leverage point, it would actually have a smaller residual than the low leverage point. Thus, not only are errors in points at the ends of the line more serious in their effect, they can also be harder to detect. More about this is given in Chapter 14.

Good software will print warnings about outliers and high leverage points, but it is always a good idea to look at plots. When in doubt, try omitting the point in question to see how much the fit changes. If it changes a lot, then you will have to decide whether you trust the point or not.

The third common problem, non-constant scatter, is not visible in Figure A5. If it were, the residuals would fan out from the line, with noticeably more variability at one end, usually the upper end, than at the other. When measurement errors increase in variability with the size of the measurement, as they often do, this pattern may be seen in the residuals. It is a potential problem because the least squares fitting criterion may not be the best in this case. It gives all the residuals equal weight in assessing the fit. If some $y$s have larger error than others then it would be more appropriate to be use a weighted version of least squares [see, for example, Weisberg (1985)]. Sometimes taking logarithms of the *y*-variable can solve this problem more simply, because proportional errors on the original scale become additive ones after taking logs.

### A.3.5 Assessing the fit numerically

To understand the common numerical measures of goodness of fit the ideas of variance and standard deviation as measures of scatter are needed.

For any set of numbers $y_1, y_2,...,y_N$, and in particular for the *y*s in the regression, the variance $s_y^2$ is defined (as in A22) as

$$s_y^2 = \frac{\sum_{i=1}^{N}(y_i - \bar{y})^2}{N-1} \quad (A32)$$

One numerical way to summarise the fit of the line is through an average squared residual. What is usually reported is the root mean square average residual

$$\hat{\sigma} = \sqrt{\frac{\sum_{i=1}^{N} r_i^2}{N-2}} \quad (A33)$$

This is called the residual standard deviation, and $\hat{\sigma}^2$ is called the residual variance. The divisor of $N-2$ is because the residuals are measured from a straight line with two parameters $b_0$ and $b_1$ fitted to the data. Dividing by $N-2$ rather than $N-1$ corrects for the fitting of the regression line, and means that $\hat{\sigma}^2$ correctly (or to be more precise, unbiasedly) estimates $\sigma^2$, the variance of the errors *f*. The residual standard deviation $\hat{\sigma}$ is a typical deviation of a data point from the fitted line, and is thus a directly interpretable measure of fit. For Forbes's data, the residual standard deviation is 0.23 inches of mercury. If we used the equation to predict pressure we would expect a typical error to be of this size, and most of them to be in the range $\pm 2\hat{\sigma}$, or roughly $\pm$ half an inch of mercury.

A second measure is the correlation coefficient $r$ between $y$ and $\hat{y}$, which varies between $-1$, for a perfectly fitting line with a negative slope, through 0, for a horizontal line, to $+1$, for a perfectly fitting line with a positive slope. There is a useful interpretation for the squared correlation between $y$ and $\hat{y}$, usually denoted by $R^2$ and called R-squared; it is the proportion of variance in the $y$s that is explained by the fitted line. In fact

$$\hat{\sigma}^2 \approx (1-R^2)s_y^2 \qquad (A34)$$

with only a missing factor of $(N-2)/(N-1)$ on the left-hand side of the equation preventing the $\approx$ being an $=$. Thus $(1-R^2)$ is the ratio of the variance about the line to the total variance in the $y$s. The smaller this is, i.e. the closer $R^2$ is to 1, the more impressive the fit is.

There is a danger of over interpreting $R^2$. The line fitted to Forbes's data has an $R^2$ of 0.994, i.e. it explains 99.4% of the variability in the pressures, 99.7% if the outlier is removed. It is tempting to think that this demonstrates the correctness of the straight line as a model for the relationship between the two measurements. However, the residual plot in Figure A5 clearly shows that the straight line is not correct. High $R^2$ only means that the fitted line is good at explaining the $y$-variable over the range studied, and that is all. It is also worth pointing out that $R^2$ depends on the range of data in the experiment. It is possible to improve $R^2$ by increasing the ranges of $x$ and $y$, and thus increasing $s_y^2$. If the straight line continues to hold over the wider range the residual variance $\hat{\sigma}^2$ will be similar over the wider range and so the larger $s_y^2$ results in a larger $R^2$.

Another problem with the $R^2$ occurs when the number of objects is small compared to the number of parameters to be fitted. In such cases, it may give over optimistic results due to random noise in the experi-

**A.3** Simple linear regression

ment. In such cases proper validation on new data is required (see Chapter 13).

In plotting and fitting a line to Forbes's data we took, without comment, pressure as the *y* variable and boiling point as *x*. The rationale for this was that we wished to predict pressure from boiling point. Doubtless it occurred to some readers that since there is a causal relationship in the opposite direction, from pressure to boiling point, it would be more logical to plot and fit the other way round.

Because least squares minimises errors in one particular direction it *does* matter which variable is *y* and which is *x*. Fortunately it often matters very little, but occasionally it can cause problems. See Section 3.2 for further discussion.

### A.3.6 Matrix form of the regression model

When there is only one predictor variable it is not necessary to use matrices either to describe the regression model or to implement the calculations. However, it may be useful for readers not familiar with the notation to see how simple regression is set up in matrix form. Writing out the model (A28) explicitly for each observation we have

$$y_1 = b_0 + b_1 x_1 + f_1$$
$$y_2 = b_0 + b_1 x_2 + f_2$$
$$\vdots \quad \vdots \quad \vdots \quad \vdots$$
$$y_N = b_0 + b_1 x_N + f_N \tag{A35}$$

It is this array of equations that is put into matrix form. We need to define three vectors,

$$\mathbf{y} = \begin{pmatrix} y_1 \\ y_2 \\ \vdots \\ y_N \end{pmatrix}, \quad \mathbf{f} = \begin{pmatrix} f_1 \\ f_2 \\ \vdots \\ f_N \end{pmatrix} \text{ and } \mathbf{b} = \begin{pmatrix} b_0 \\ b_1 \end{pmatrix}$$

and one $N \times 2$ matrix

$$\mathbf{X} = \begin{pmatrix} 1 & x_1 \\ 1 & x_2 \\ \vdots & \vdots \\ 1 & x_N \end{pmatrix}$$

Then we can write the array of equations rather more concisely as

$$\mathbf{y} = \mathbf{X}\mathbf{b} + \mathbf{f} \tag{A36}$$

The arrays of $y_i$ and $f_i$ have simply been replaced by the symbols $\mathbf{y}$ and $\mathbf{f}$. The other part is more complicated as it involves a matrix multiplication. The result of multiplying the $N \times 2$ matrix $\mathbf{X}$ by the $2 \times 1$ vector $\mathbf{b}$ is an $N \times 1$ vector whose $i$th element is $b_0 + b_1 x_i$, obtained by multiplying the $i$th row of $\mathbf{X}$ into $\mathbf{b}$ (see Appendix A1 for more details of matrix multiplication, and of the transposition and inversion that will be used below). The reason for setting up the matrix representation in this particular way is that it separates the predictor variables, in $\mathbf{X}$, from their coefficients, in $\mathbf{b}$, so that each symbol in the equation represents one of the four types of quantity involved: observations, predictors, coefficients and errors.

The matrix notation has also allowed us to write the equations in a compact form. However, the real gain is that it enables formulae such as

$$\hat{\mathbf{b}} = (\mathbf{X}'\mathbf{X})^{-1}\mathbf{X}'\mathbf{y} \tag{A37}$$

for the least-squares estimates of the coefficients to be written just as concisely. This formula can be used for both simple and multiple regressions. In the case of simple regression it is not too hard to follow through the calculations and recover a non-matrix formula that may be familiar to many readers.

First, we calculate $\mathbf{X}'\mathbf{y}$ as the matrix product of the transpose of $\mathbf{X}$ with $\mathbf{y}$

**A.3 Simple linear regression**

$$\begin{pmatrix} 1 & 1 & \cdots & 1 \\ x_1 & x_2 & \cdots & x_N \end{pmatrix} \begin{pmatrix} y_1 \\ y_2 \\ \vdots \\ y_N \end{pmatrix} = \begin{pmatrix} \Sigma y_i \\ \Sigma x_i y_i \end{pmatrix} \quad (A38)$$

where $\Sigma$ denotes summation over $i$ from 1 to $N$. Then we calculate $\mathbf{X}'\mathbf{X}$ as

$$\begin{pmatrix} 1 & 1 & \cdots & 1 \\ x_1 & x_2 & \cdots & x_N \end{pmatrix} \begin{pmatrix} 1 & x_1 \\ 1 & x_2 \\ \vdots & \vdots \\ 1 & x_N \end{pmatrix} = \begin{pmatrix} N & \Sigma x_i \\ \Sigma x_i & \Sigma x_i^2 \end{pmatrix} \quad (A39)$$

The next task is to invert this $2 \times 2$ matrix to give $(\mathbf{X}'\mathbf{X})^{-1}$. Inverting a $2 \times 2$ matrix is not particularly difficult in general, but it is even easier if the off-diagonal (i.e. the top right and bottom left) elements are zero. We can arrange for this to be so if we centre the $x$-values in advance by subtracting the mean from each of them. Then their sum is zero so that using centred $x$ we have

$$(\mathbf{X}'\mathbf{X})^{-1} = \begin{pmatrix} N & 0 \\ 0 & \Sigma x_i^2 \end{pmatrix}^{-1} = \begin{pmatrix} 1/N & 0 \\ 0 & 1/\Sigma x_i^2 \end{pmatrix} \quad (A40)$$

Beware: only diagonal matrices are inverted by taking one over the individual elements like this, the general rule is more complicated. Finally, we can multiply this inverse into $\mathbf{X}'\mathbf{y}$ to give

$$\begin{pmatrix} \hat{b}_0 \\ \hat{b}_1 \end{pmatrix} = \hat{\mathbf{b}} = (\mathbf{X}'\mathbf{X})^{-1}\mathbf{X}'\mathbf{y}$$

$$= \begin{pmatrix} 1/N & 0 \\ 0 & 1/\Sigma x_i^2 \end{pmatrix} \begin{pmatrix} \Sigma y_i \\ \Sigma x_i y_i \end{pmatrix} = \begin{pmatrix} \Sigma y_i / N \\ \Sigma x_i y_i / \Sigma x_i^2 \end{pmatrix} \quad (A41)$$

so that the least-squares intercept $\hat{b}_0$ for centred $x$ is the mean $y$ and the least-squares slope $\hat{b}_1$ is the sum of products of (centred) $x$ and $y$ divided by the sum of squares of (centred) $x$. These well-known results could of course be derived without the use of matrices. The matrix forms, however, generalise to more than one $x$.

## A.4 Multiple linear regression (MLR)

When there are several explanatory variables $x$ to be used for predicting $y$, one needs multiple linear regression (MLR), which is a straightforward generalisation of the linear regression theory above

The model used is identical, except that the number of regressors is increased from 1 to $K$. The model can be written as:

$$y = b_0 + \sum_{k=1}^{K} b_k x_k + f \qquad (A42)$$

As before, the regression coefficients are considered fixed unknown constants (parameters) and the error $f$ is considered random. The expectation of $f$ is assumed to be 0 and its variance is usually denoted by $\sigma^2$ and is considered to be the same for each observation.

We can write this model in exactly the same matrix form

$$\mathbf{y} = \mathbf{Xb} + \mathbf{f} \qquad (A43)$$

as (A36) if we define $\mathbf{y}$ and $\mathbf{f}$ as before,

$$\mathbf{X} = \begin{pmatrix} 1 & x_{11} & \cdots & x_{1K} \\ 1 & x_{21} & \cdots & x_{2K} \\ \vdots & \vdots & \ddots & \vdots \\ 1 & x_{N1} & \cdots & x_{NK} \end{pmatrix} \qquad (A44)$$

where $x_{ij}$ is the value of the $j$th predictor for the $i$th case, and

$$\mathbf{b} = \begin{pmatrix} b_0 \\ b_1 \\ \vdots \\ b_K \end{pmatrix} \quad (A45)$$

For MLR, the equation is usually fitted to the data using the same criterion as for simple regression, namely least squares. For a given **b** we can calculate fitted values using the equation, subtract these from the observations **y** to get residuals, and assess the fit by summing the squares of the $N$ residuals. Least squares chooses the value $\hat{\mathbf{b}}$ of **b** that minimises this sum of squares. This problem can be solved mathematically, by differentiation and setting the derivatives equal to 0 for example. The solution

$$\hat{\mathbf{b}} = (\mathbf{X}'\mathbf{X})^{-1}\mathbf{X}'\mathbf{y} \quad (A46)$$

has already been seen in (A37). With more than one predictor it is not useful to try to derive formulae for the individual coefficients as we did above. Instead the calculations may be programmed directly in matrix form.

If each of the predictor variables has been centred in advance by subtracting the mean over the $N$ cases, it is possible to separate the estimation of $b_0$ (by the mean $y$) from that of the other $K$ coefficients. To estimate these we still use formula (A46), but with the leading column of ones deleted from **X** (which now has the centred $x$ in its other columns) and the leading $b_0$ deleted from **b**.

The statistical distribution of the regression parameter estimate vector is very easy to derive. If the original variables $y$ are normally distributed, $\hat{\mathbf{b}}$ will be multivariate normal as well. It is always unbiased, with covariance matrix equal to

$$\text{cov}(\mathbf{b}) = (\mathbf{X}'\mathbf{X})^{-1}\sigma^2 \qquad (A47)$$

regardless of what distributional assumptions are made about $\mathbf{y}$.

For the matrix inversion in A46 and A47 to be possible the matrix $\mathbf{X}'\mathbf{X}$ has to satisfy certain conditions. If it does it is described as non-singular or of full rank. A necessary condition for this to hold is that $N$ is greater than or equal to $K + 1$, that is we must have at least as many samples as variables. Even if this condition is satisfied we may still run into problems if the predictor variables are very strongly correlated with each other or, even worse, one predictor variable can be expressed exactly as a linear combination of some of the others. In this latter case we say that we have exact collinearity, and $\mathbf{X}'\mathbf{X}$ is singular. Then the estimate of $\mathbf{b}$ does not exist uniquely. With strong correlations between the predictor variables but no exact relationship, the matrix may have an inverse in principle, but its computation may be highly numerically unstable. One can, however, often obtain good prediction results in both these situations if a method other than least squares is used for fitting. This is one of the main topics in Chapter 5 of this book.

In two dimensions, the least-squares regression is easy to visualise graphically. If $y$ is the vertical axis and the two $x$-variables are represented in the horizontal plane, the fitting of linear equation by LS is identical to finding a plane that fits the best possible way to the data, when fit is measured by the sum of squares of the residuals, i.e. the vertical distances from the data points to the plane.

For MLR we can use similar numerical measures of fit to those described above for simple regression. The residuals, denoted by $\hat{f}_i$, are defined as the differences between the predicted and measured $y$-values, i.e. $\hat{f}_i = y_i - \hat{y}_i$. As in the simple regression case, the sum of squares of the residuals is denoted by

*RSS*. An unbiased estimator for the error variance $\sigma^2$ is calculated as

$$\hat{\sigma}^2 = RSS/(N-1-K) \qquad (A48)$$

The multiple correlation coefficient $R^2$ can be defined as the square of the simple correlation between the measured and predicted *y*-values. As in simple regression $R^2$ can be interpreted as the proportion of *y*-variance explained by the predictor variables.

The aspect of two regression "lines" as was described above generalises to several dimensions. The so-called classical approach will, however, not be pursued here. The main reason for that is that we generally think that the methods using *y* on the left side of the equation are usually better and easier to use than the other methods. For more information, see, for example, Martens and Næs (1989) and Section 3.2.

Significance tests for the regression coefficients can also easily be developed if the error terms are assumed to be the normally distributed. This can be done in various ways, but the easiest is probably the t-test for each of the regression coefficients separately. These are obtained by just dividing the coefficient values by the corresponding standard deviations obtained from equation (A47). The error variance must be estimated by the estimate in equation (A48). The corresponding quantity will have a Student t-distribution with $N - K - 1$ degrees of freedom. The square of this will have an F-distribution with 1 and $N - K - 1$ degrees of freedom. This is exactly the F-test used for variable selection in the stepwise regression approach (see Chapter 7).

For MLR it is also possible to derive an estimate of the variability of the prediction errors. To predict the value of *y* corresponding to an observed vector **x**, the appropriate formula is

$$\hat{y} = \mathbf{x}'\hat{\mathbf{b}} \qquad (A49)$$

This prediction has variance given by

$$\text{var}(\hat{y}) = \sigma^2 \mathbf{x}^t (\mathbf{X}^t \mathbf{X})^{-1} \mathbf{x} \quad (A50)$$

where $\sigma^2$ is replaced by its estimate when necessary. This error variance only considers the error in the prediction process, and does not take account of the random error in the $y$ that is actually observed. If we compare our predictions with actual observations, themselves having variance $\sigma^2$, the total variance for the differences needs to be increased to

$$\sigma^2 (1 + \mathbf{x}^t (\mathbf{X}^t \mathbf{X})^{-1} \mathbf{x}) \quad (A51)$$

All the above theory was developed assuming that the error terms are independent, identically distributed with the same variance. The methodology can, however, easily be generalised to situations with a more complicated error structure. For instance, if the residuals have different variance for the different observations, and the relative size of the variance are known, the LS theory can easily be modified. This is called weighted least squares [WLS, see Weisberg (1985)].

> Note that var($\hat{y}$) depends on **x**. This is different from the prediction errors discussed in Chapter 13. In Chapter 13, the main emphasis is put on average error estimates, where the average is taken over the test samples.

## A.5 An intuitive description of principal component analysis (PCA)

Principal component analysis (PCA) is a method of data reduction or data compression. Applied to a data matrix of samples by variables it constructs new variables, known as principal components (PCs), such that:

- the new variables are linear combinations of the original ones, with the weight vectors that define the combinations being of unit length and orthogonal (at right angles) to each other;
- the new variables are uncorrelated;
- the first new variable captures as much as possible of the variability in all the original

ones, having been constructed to have maximum variance amongst all such linear combinations;
▸▸ each successive new variable accounts for as much of the remaining variability as possible.

The reason the method is useful is that it is often possible to describe a very large proportion of the variability in highly multivariate data with a modest number of these new variables. This helps with visualisation, a scatterplot of the first two new variables often being highly informative, but more importantly the first few new variables can be used as input to methods such as multiple regression or discriminant analysis that will not cope with highly multivariate data.

Mathematically, the construction of the new variables is achieved by finding the eigenvectors (see equation A20) of the variance matrix of the original variables. These eigenvectors then become the weight vectors for construction of the variables, and the corresponding eigenvalues tell how much of the original variance has been captured in each new variable. This is, alas, rather technical, but the transformation from old to new variables is simply a rotation of axes and is easy to visualise geometrically. The pictures below are for the case of three original variables, because three-dimensional spaces are easy to draw. The mathematics does not change as the number of dimensions increases, the matrices just get larger. If you can see what is happening in these pictures, you have understood the general case.

Figure A6 shows a three-dimensional plot for three variables measured on a set of 72 samples. Three of the samples are specially marked to aid visualisation of the changes, they are not special samples. In PCA we normally begin by centring the data so that it varies around zero. This is known as "mean centring" as it is done by subtracting a mean value. For each

**A.5** An intuitive description of principal component analysis (PCA)

# Appendix A. Technical details

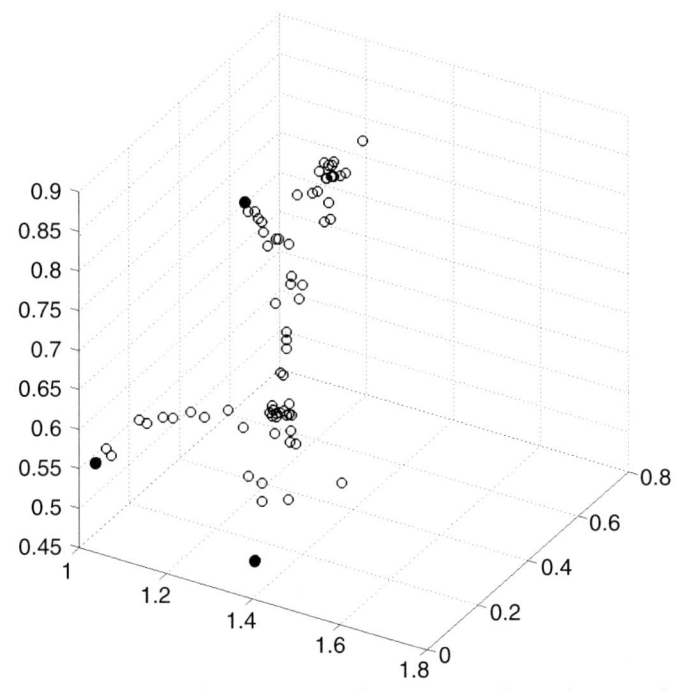

**Figure A6.** Data from 72 samples measured on three variables.

variable we compute the mean over all samples, and then subtract this value from each measurement of that variable. This operation does not change the shape of the cloud of points that represent the data; it just shifts it, as can be seen in Figure A7 which is the same data after mean centring. Now the origin of the three axes is in the middle of the cloud of points.

Now we find the first new variable, i.e. the first PC. The axis for this variable, shown in Figure A8, is in the direction in which the three-dimensional data show the maximum amount of variability as measured by their variance. It has its origin in the same place as the original axes. The value of the new variable for any particular sample is found by projecting the point for the sample onto this axis (see equation A19), i.e. by finding the point on the axis nearest to the point

**A.5** An intuitive description of principal component analysis (PCA)

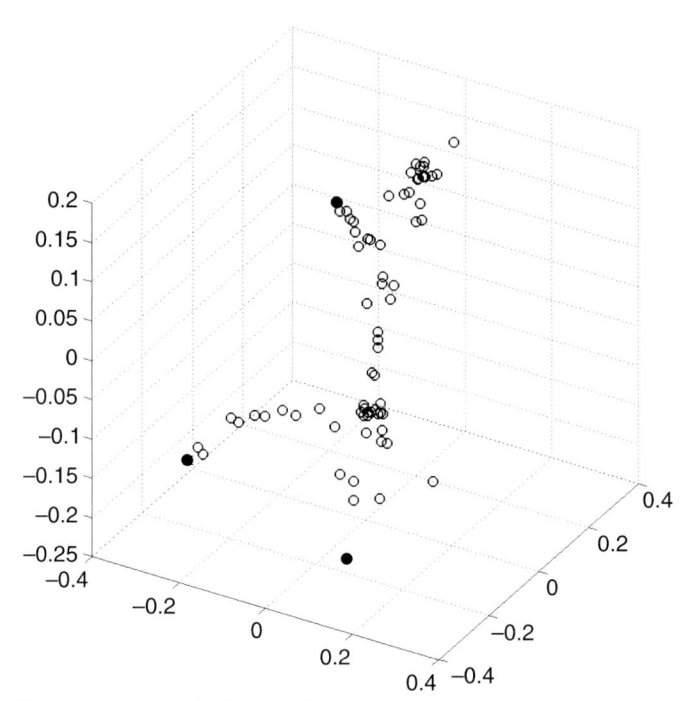

**Figure A7.** Data in Figure A6 after mean centring.

representing that sample in the three-dimensional space. This is the point at the foot of a perpendicular line joining the point to the axis. These values are called the scores of the samples on the first PC.

In Figure A9, the axis for the second PC has been added. This also passes through the origin, is at right angles to the first PC and is again oriented in the direction in which there is maximum variability in the data points. Imagine spinning the first axis so that this second one produces a circular disc. The orthogonality constraint restricts the second axis to lie in this disc. Given this constraint, it is chosen to capture as much variability as possible. As before, the scores may be found by projecting the original points onto this new axis.

In Figure A10 the third and final PC has been added. Since it must be at right angles to both of the

# Appendix A. Technical details

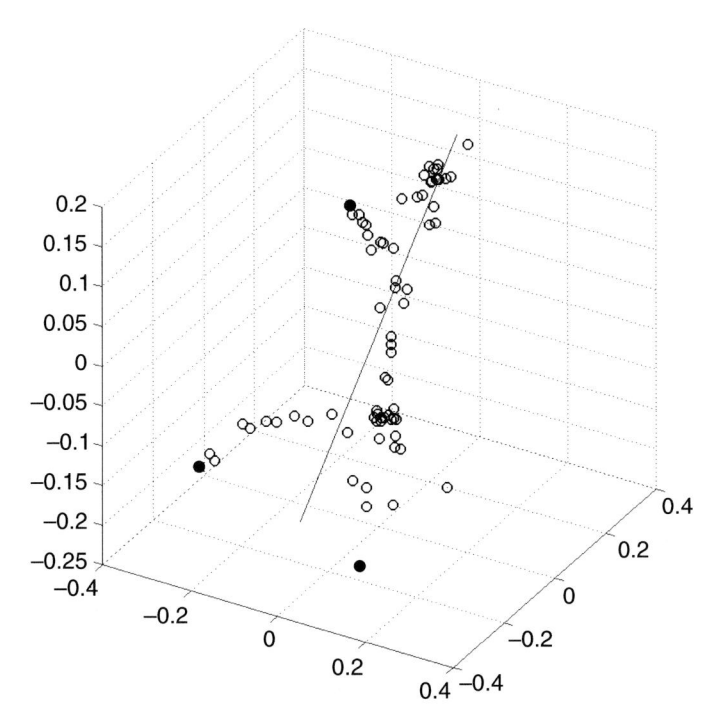

**Figure A8.** First principal component (PC) plotted in the three-dimensional data space.

first two there is only one possible direction. It captures whatever variability in the data points is left over. Now we have a new system of three orthogonal axes, with the same origin as the original ones. Thus one can see the transformation to principal components geometrically as a rotation of the original axes into an ordered set of new axes so that the first describes as much as possible of the variation in the data as is possible with one variable, the first two describe as much as is possible with two, and so on.

The whole point of this rotation is that we can obtain good lower-dimensional representations of the data by discarding all but the first few PCs. In an example where there were 100 variables to start with, one might rotate into 100 new axes and discard all ex-

**A.5** An intuitive description of principal component analysis (PCA)

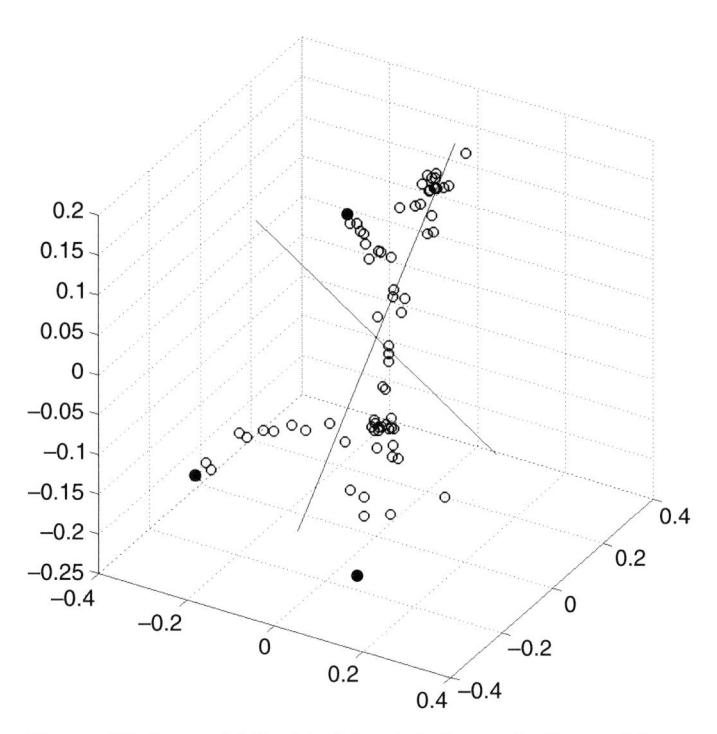

Figure A9. Second PC added to plot shown in Figure A8.

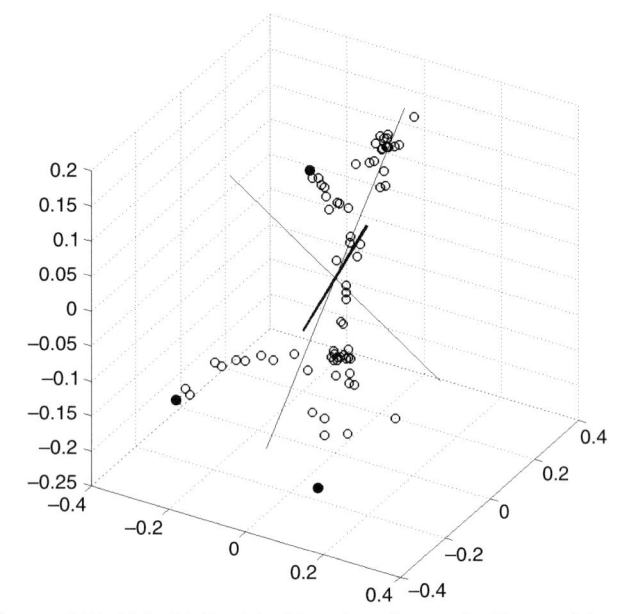

Figure A10. Third PC added to plot shown in Figure A9.

**A.5** An intuitive description of principal component analysis (PCA)

cept the first ten, say. In this smaller example we can discard the third PC and still get a good picture of the variability in the data. To do this we project the points onto the plane defined by the first two axes, resulting in the scores plot in Figure A11. Imagine looking at the cloud of points in Figure A10 with the third PC axis pointing straight at your eyes. Figure A11 is what you would see. In this sense PCA has given us a two-dimensional window on the three-dimensional data, chosen so that what we see is as interesting as possible.

Even when we start with much larger number of variables, this two-dimensional window on the data can be highly informative. Sometimes we need to supplement it with plots of a few higher-order PCs to get the whole picture. When the PCA is a preliminary data-reduction step for multiple regression we will usually retain a larger number of PCs, perhaps choosing this number by optimising some measure of predictive performance. Fortunately, it is often the case

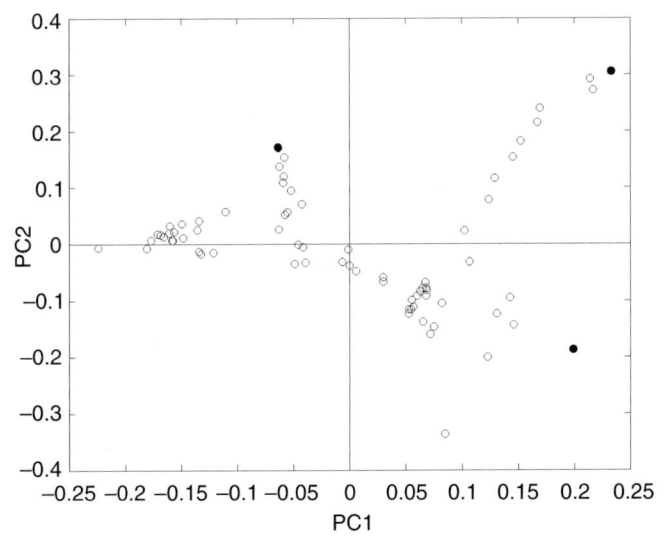

**Figure A11. Plot of scores for the 72 samples on the 1$^{st}$ and 2$^{nd}$ PCs.**

**A.5** An intuitive description of principal component analysis (PCA)

that, even when the original number of variables is very large, a modest number of PCs captures essentially all of the variability, and hence all of the information, in the data.

The sum of the variances for the original variables is equal to the sum of the variances of the score values. Therefore it is meaningful to talk about percent explained variance for each of the components. The components as extracted above will always have decreasing variance values. Often in NIR analysis, a very large percent (often more than 95%) of the total variance is explained by the first few (2–3) principal components. This means that most of the variability in the spectrum can be compressed to a few principal components with only a rather small loss of information.

# B Appendix B. NIR spectroscopy

## B.1 General aspects

Data from NIR (near infrared) spectroscopy have been a catalyst for the developments and applications of many of the data analysis methods used in this book. This chapter gives a brief description of the basics of the technique. For more information see Osborne *et al.* (1993) and Williams and Norris (1987).

NIR spectroscopy is a fast, low-cost, easy-to-use, non-destructive, reliable and versatile analytical method. NIR penetrates deep into the sample, which is important for heterogeneous samples. Moisture is a relatively low absorber in NIR, which is also important when measuring biological products. NIR is today much used in determining a variety of quality parameters in a large number of biological materials, such as grains, feeds, meat and meat products, vegetables and fruits. It is used both in the laboratory and as an on-line analysis technique.

The discovery of invisible radiation was made by Herschel (1800) when he observed that the relative heating effect of different portions of the spectrum of sunlight generated by a prism continued to increase when he placed a thermometer beyond the red end of the visible spectrum. He had discovered what is now known as near infrared radiation.

The NIR region of the electromagnetic spectrum was little used until late 1950. The main reason for that was its complexity, compared to the so-called mid- and far-infrared (MIR and FIR) regions. MIR and FIR spectra give much useful information by using univariate approaches, i.e. the spectroscopic peaks are resolved and can be interpreted directly. The NIR spectra are results of many overlapping and often broad peaks, which are difficult or impossible to interpret using univariate methods. A NIR spectrum is a result of overtone absorbances and combinations of absorbances from several functional groups, such as C–H, N–H and O–H.

In addition, a diffuse reflectance or transmittance NIR spectrum is a result of the physical conditions of both the instrument and the sample: instrument geometry; sample particle and droplet size, form and distributions; refractive indices etc. If one measures biological products, it is impossible (or very time consuming) to measure without diffuse scattering of the light. This would need a full purification of the samples, and important interaction and synergy effects between the constituents would be lost. The resulting NIR spectrum from a diffuse measurement is therefore a mixture of several physical and chemical effects. This is the reason why extensive use of multivariate data analytical methods is necessary to reveal specific and useful information from such NIR spectra.

NIR spectra are usually presented in "absorbance" units defined by either $A = \log(1/R)$ or $A = \log(1/T)$, depending on whether the data are based on reflectance ($R$) or transmission ($T$).

Examples of NIR spectra are presented in several places in the book, for instance Figure 3.3 and Figure 6.4. These represent two different parts of the region.

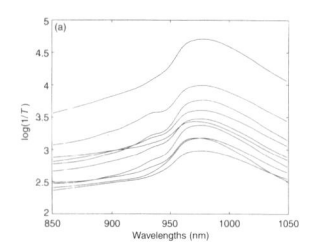

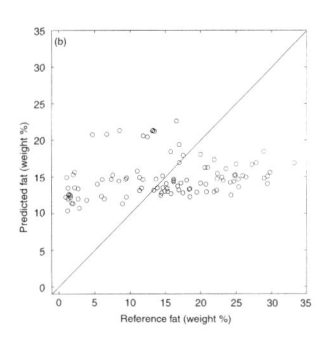

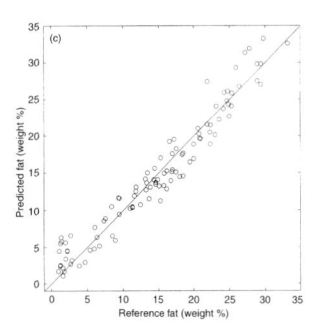

**Figure 3.3.** Illustration of the selectivity problem in NIR spectroscopy. Pork and beef data.

These were defined by IUPAC in 1984.

## B.2 More specific aspects

The NIR region of the electromagnetic spectrum extends from 780 to 2500 nm (wavelength) or 12,800 to 4000 cm$^{-1}$ if measuring in wavenumbers (the number of waves per cm). NIR spectroscopy is concerned with absorptions of NIR energy by molecules within a sample. The absorption of energy increases the energy of a vibration in a molecule. Although most NIR spectroscopists think in terms of wavelength it is easier to follow the theory if we think in wavenumbers.

Absorptions in this region are caused by three different mechanisms. These are:

▸▸ Overtones of fundamental vibrations which occur in the infrared region (4000–300 cm$^{-1}$)
▸▸ Combinations of fundamental vibrations which occur in the infrared region
▸▸ Electronic absorptions

### B.2.1 Overtones

Overtones are approximately multiples of the fundamental vibrations. A fundamental vibration, "f", will give rise to a series of absorptions 2f, 3f, 4f, ... which are known as the first, second, third overtone. The intensity of these successive absorptions will decrease by a factor between 10 and 100.

### B.2.2 Combinations

The absorption process is concerned with photons. In overtones, a photon of the appropriate (NIR) energy is required to raise the energy of a vibration to the higher energy state i.e. first, second, third overtone. In a combination, absorption a photon of NIR energy is shared between two (or more) vibrations which would be individually observed as fundamentals in the MIR region. If we have fundamental absorptions f1 and f2 which occur at 3000 and 1600 cm$^{-1}$, then these would give rise to a combination band at approximately 3000 + 1600 = 4600 cm$^{-1}$. This translates to 2174 nm, which is an NIR absorption. Combinations can be multiples of one or more of the vibrations and there is no theoretical limit to the number of absorptions that can be involved. Thus our two vibrations f1 and f2 can give rise to a myriad of absorptions for example: $n$f1 + $m$f2 ($n$ = 1, 2, 3, 4, ...;

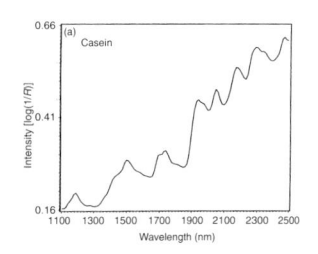

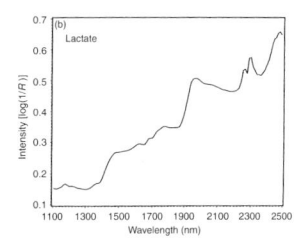

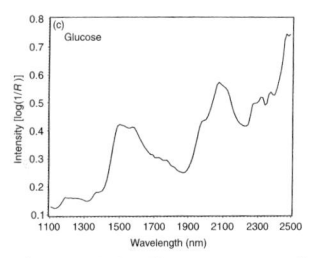

Figure 6.4. The spectra of the three pure ingredients (a) casein, (b) lactate and (c) glucose in the experiment.

**B.2** More specific aspects

$m = 1, 2, 3, 4, \ldots$). If we involve a third vibration, f3, then the number of possibilities becomes very large. Combination bands have two effects on NIR spectra:
- Absorptions occur at unexpected positions in the NIR spectrum
- Regions of absorption occur as broad peaks caused by the overlapping of a multitude of different absorptions

### B.2.3 Electronic absorptions

Electronic absorptions are caused by the movement of electrons from one orbit to a higher-energy orbit. They are normally observed in the visible or ultraviolet regions of the electromagnetic spectrum but they can also occur in the NIR region, especially in the 780–1100 nm region. The most useful of these are in the spectra of rare earths.

The consequence of these different mechanisms makes an NIR spectrum very complex but often this complexity is hidden as a few rather broad areas of absorption.

### B.2.4 Hydrogen bonding

Hydrogen bonding is caused by the tendency of hydrogen to form weak bonds with electron donating atoms, especially oxygen and nitrogen. The formation of a weak bond will affect all the vibrations with which that hydrogen and its electron-donating partner are associated. This results in peak shifts (to longer, less energetic wavelengths) which are often observed as peak broadening. The hydrogen bonding structure of water is very complex and depends on temperature, pH, ionic concentration etc., so whenever water is present in a sample there will a complex, dynamic interaction between water and the sample, which may be unique for that sample.

Although NIR spectra are very complex, the fact that the same atoms are involved in many different absorptions means that these absorptions can be utilised, via complex mathematical analysis, to provide analytical information on specific functional groups.

# C Appendix C. Proposed calibration and classification procedures

This book covers and gives an overview of a number of different calibration and classification techniques. Although the text is meant as a guided tour through the various problems and solutions, the less experienced reader may possibly find it convenient to have an even more explicit and concisely set out route to follow in a practical and concrete application. This section is meant as a help to such users.

It is important to emphasise that these routes are only suggestions from our side and certain situations may possibly call on other strategies and ways of combining the various tools in the book. We also admit that this area is closely related to personal preferences and working habits. No clear absolutely correct route can therefore be set up.

## C.1 Multivariate calibration

1. The first step in a calibration process should always be to define the problem as clearly as possible. This should focus on what kind of measurements ($\mathbf{x}$,$y$) to take and which population of samples to consider. Signal-to-noise ratios, precision, selectivity and other measurement characteristics should also be given attention. Of course, these measurement characteristics should be as

good as possible, so fine-tuning of instruments may be an important part of this first step. As much information about these issues as possible should be present prior to calibration, to help interpretation and safe use of the techniques.
2. Calibration samples should then be selected. For this phase, one needs to check whether the situation of interest fits naturally into one of the examples given in Section 15.1. If this is true, these ideas and principles can be followed directly. In extreme cases with little prior information, the only chance may be to take random samples from the target population, but this is risky. Therefore one should try to follow the principles in Chapter 15 as closely as possible
3. As soon as samples are selected, measurements of reference values and spectral data can be made. The calibration data should first be checked using simple plotting techniques to look for obvious mistakes and misprints. The plotting of spectral absorbances against wavelength is one such technique. Range of variability (max and min), mean and standard deviations should also be computed. The same is true for correlations and the condition number to check for collinearity.
4 Time has now come for the calibration itself, the model building phase. This may be a time-consuming exercise if one is unlucky, i.e. if the relationship between $y$ and $x$ is complex.
(a) The first thing one should do, especially if the collinearity is substantial, is usually to run PLS or PCR and check their results. Standardisations should also be considered at this point (see Section 5.2). This checking of results should involve looking at the plots (Chapter 6), outlier diagnostics (Chapter 14) and a test set validation or cross-validation (Chapter 13). If everything looks OK,

one can of course stop and go to step 5. If not, one can do several things.
(b) If there are clear outliers in the data set, these should be taken seriously and handled according to the guidelines in Section 14.3.
(c) If the residual plot in (a) indicates non-linearities or if the calibration is not good enough, one should try one of the remedies discussed in Chapter 9 and treated more thoroughly in the subsequent chapters (Chapters 10, 11 and 12). Which one to use is somewhat difficult to decide and is also a matter of preference. Our preference would be to try one of the scatter correction techniques (for instance MSC, Chapter 10) followed by a LWR (Chapter 11). These are methods which have shown good performance and which are also easy to run, interpret and use. They are, however, not necessarily optimal. A careful study of the possibility of deleting some of the variables is also something which could be conducted at this point (Chapter 7, if PLS is used the methods in 7.4 are useful).
(d) If the interpretation of plots is difficult or the number of components needed is very large, this may also be an argument for trying the methods in (c) above. A scatter correction may reduce the number of components substantially and non-linear modelling can sometimes be used. The data may also become easier to interpret after a scatter correction.
(e) Generally, a calibration is not a one-shot exercise. It often needs some trial and error and some iterations between modelling, variable selection, interpretation, outlier detection and validation. In particular, this is true in situations with limited prior knowledge about the application under study. The more is known, the easier is of course the calibration.

**C.1 Multivariate calibration**

5. The validation should always be taken seriously regardless of which of the methods in step 4 is eventually chosen. The methods in Chapter 13 are good candidates. They are applicable for any calibration method and easy to run in practice. In a thorough validation process, possible outliers (Chapter 14) and model quality (Section 9.2) should preferably also be considered. A plot of the measured and predicted $y$ for the validation objects is very useful at this point.
6. When a calibration is developed and one is satisfied and confident with the results, the calibration can be put into production. Before one relies too heavily on it, one should, however, test it carefully in a practical environment. This means that frequent tests using the monitoring methods described in Chapter 16 should be used. As one gains confidence in a calibration, the less frequent the tests need to be. They should, however, never be forgotten completely.
7. If a clear drift or change takes place during regular operation one can use one of the methods in Chapter 17 for correction. Continuous updating of slowly drifting systems can also be envisioned [see Helland *et al.* (1992)].

## C.2 Multivariate discriminant analysis

For discriminant analysis, steps 1–4 are almost identical, the only difference being that the selection of samples (Chapter 15) and checking for outliers (Chapter 14) should be done for each group of samples.

When comes to discriminant analysis itself, one has a number of different ways to go. A simple way, which is easy to run on most statistics packages, is to use LDA or QDA based on principal components (Section 18.5.1). If the data are not collinear, the LDA or QDA can be used on the raw data directly. This is a rather robust and safe strategy, but it is not necessarily optimal. One of the advantages of this method is that it is based on two very simple and established steps, which are both well understood theoretically. It also provides simple plotting tools from the PCA. LDA should be tried first because of it simplicity and stability. Another strategy, which is closely related, is to use discriminant PLS (Section 18.6.2) followed by a LDA or QDA based on the computed PLS components. If groups have very irregular shapes, KNN (Section 18.6.1) based on principal components may be a good alternative. If one is in doubt about which method to use or one is not satisfied with the results obtained, one can run a number of the methods discussed in Chapter 18 and compare the results by cross-validation or prediction testing.

Validation (Chapter 13) is equally important here. Pre-processing (for instance, using MSC, Chapter 10) may also be advantageous to improve interpretability and classification ability. Updating and standardisation (Chapter 17) and monitoring (Chapter 16) can be done the same way as for calibration.

## C.3 Cluster analysis

Since cluster analysis is highly exploratory, it is more difficult to set up a structured method of analysis. A reasonable first attempt is to first compute the principal components, then compute the truncated Mahalanobis distances based on the most dominating distances and use complete linkage accompanied with an interpretation of the dendrogram (Section 18.9.2). PCA and computed means for the different groups can be used to help interpretation.

# Subject index

## A

Absorbance . . . . . . . . . . . . . . . . . . . . . . . . 106
Accuracy . . . . . . . . . . . . . . . . . . . . . . . . . 164
Additive effect . . . . . . . . . . . . . . . . . . . . . 106
Akaike information criterion . . . . . . . . . . . . . 58
ANOVA, Analysis of variance . . . . . . . . . . . . 167

## B

Backward elimination . . . . . . . . . . . . . . . 60, 61
Beer's law . . . . . . . . . . . . . . . . 12, 44, 47, 98
Best subset selection . . . . . . . . . . . . . . . 24, 60
Bias . . . . . . . . . . . . . . . . . . . . . . . . . . . 164
    correction . . . . . . . . . . . . . . . . . . . . 217
Bootstrap . . . . . . . . . . . . . . . . . . . . . . . . 162
Box–Cox transformation . . . . . . . . . . . . . . . 100

## C

Calibration . . . . . . . . . . . . . . . . . . . . . . . . . 5
    classical . . . . . . . . . . . . . . . . . . . . . . 11
    inverse . . . . . . . . . . . . . . . . . . . . . . . 11
    multivariate . . . . . . . . . . . . . . . . . . 5, 329
    outlier . . . . . . . . . . . . . . . . . . . . . . . 178
    robust . . . . . . . . . . . . . . . . . . . . . . . 216
    transfer . . . . . . . . . . . . . . . . . . . . . . 209
    univariate . . . . . . . . . . . . . . . . . . . . . 11
Classical calibration . . . . . . . . . . . . . . . . . . 11
Classification . . . . . . . . . . . . . . . . . . . . . . . 5
    multivariate . . . . . . . . . . . . . . . . . . . . . 5
    supervised (discriminant analysis) . . . . . 221, 222, 332
    unsupervised (cluster analysis) . . . . . . . 221, 249, 333
Cluster analysis . . . . . . . . . . 5, 142, 197, 221, 249, 333
    complete linkage . . . . . . . . . . . . . . . . 251
    fuzzy clustering . . . . . . . . . . . . . . . 142, 256
    hierarchical methods . . . . . . . . . . . . 142, 251

membership value . . . . . . . . . . . . . . . . . . . . 256
partitioning methods . . . . . . . . . . . . . . . . . . 255
single linkage . . . . . . . . . . . . . . . . . . . . . . 251
using PCA . . . . . . . . . . . . . . . . . . . . . . . . 249
Collinearity . . . . . . . . . . . . . . . . . . . . . . . . . . . 6, 19
see also Multicollinearity
Component . . . . . . . . . . . . . . . . . . . . . . . . . . . . 39
see also Principal component
Condition number (CN) . . . . . . . . . . . . . . . . . . 21, 35
Confidence intervals . . . . . . . . . . . . . . . . . . . . 170, 171
for predictions . . . . . . . . . . . . . . . . . . . . . . 171
Correlation (coefficient) . . . . . . . . . . . . . . . . . 297, 307
multiple correlation coefficient (R-squared) . . . . . . 314
Covariance (matrix) . . . . . . . . 21, 227, 234, 295, 298, 313
Cross-validation (CV) . . . . . . . . . . . . . . . . 15, 59, 160
full . . . . . . . . . . . . . . . . . . . . . . . . . . . . . 161
segmented . . . . . . . . . . . . . . . . . . . . . . . . 161
Continuum regression . . . . . . . . . . . . . . . . . . . . . 37

# D

Data compression . . . . . . . . . . . . . . . . . . . . . . . . 22
by PCR and PLS . . . . . . . . . . . . . . . . . . . . . 27
by variable selection . . . . . . . . . . . . . . . . . . . 55
by Fourier analysis . . . . . . . . . . . . . . . . . . 71, 81
by wavelets . . . . . . . . . . . . . . . . . . . . . . 71, 84
derivative . . . . . . . . . . . . . . . . . . . . . . . 99, 107
first derivative . . . . . . . . . . . . . . . . . . . . . . 110
second derivative . . . . . . . . . . . . . . . . . . . . 110
Discriminant analysis . . . . . . . . . . . . . . . . . 5, 222, 332
Bayes rule . . . . . . . . . . . . . . . . . . . . . . . . 225
canonical variates . . . . . . . . . . . . . . . . . . 230, 231
DASCO . . . . . . . . . . . . . . . . . . . . . . . . . 240
Fisher's linear discriminant analysis . . . . . . . . . . 230
K-nearest neighbours (KNN) . . . . . . . . . . . . . . 243
linear discriminant analysis (LDA) . . . . . . . . . . 226
linear discriminant function . . . . . . . . . . . . . . 231
outliers . . . . . . . . . . . . . . . . . . . . . . . . . . 248
Prior and posterior probability . . . . . . . . . . . . 225, 226

quadratic discriminant analysis (QDA) . . . . . . . . 228
SIMCA . . . . . . . . . . . . . . . . . . . . . . . 237
using regression . . . . . . . . . . . . . . . . . . . 245
validation . . . . . . . . . . . . . . . . . . . . . . 247
Distance measure . . . . . . . . . . . . . . . . 131, 251, 256
Euclidean . . . . . . . . . . . . . . . . . . . . 132, 241
Mahalanobis . . . . . . . . . . . . . . . . . 131, 228, 237

## E

Eigenvalue . . . . . . . . . . . . . . . . . . . 21, 31, 234, 294
Eigenvector . . . . . . . . . . . . . . . . . . . 22, 31, 231, 294
Empirical covariance matrix . . . . 21, 31, 227, 234, 295, 298
Error matrix . . . . . . . . . . . . . . . . . . . . . . . . . 39
error variance . . . . . . . . . . . . . . . . . . . . 314
Euclidean distance . . . . . . . . . . . . . . . . . 132, 241
Expectation . . . . . . . . . . . . . . . . . . . . . . . . . 12
Experimental design . . . . . . . . . . . . . . . . . . . . 191
end-point design . . . . . . . . . . . . . . . . . . 196
uniform design . . . . . . . . . . . . . . . . . . . 196
selection of samples for calibration . . . . . . . . . 191
selection of samples using cluster analysis . . . . . . 197
selection of samples for calibration transfer . . . . . 212

## F

F-test . . . . . . . . . . . . . . . . . . . . . 58, 60, 186, 314
Factor . . . . . . . . . . . . . . . . . . . . . . . . . . . 39
see also Principal component analysis
Forward selection . . . . . . . . . . . . . . . . . . 24, 60, 61
Fourier analysis . . . . . . . . . . . . . . . . . . . . . 71, 73
coefficients . . . . . . . . . . . . . . . . . . . . 79, 81
cosine (waves) . . . . . . . . . . . . . . . . . . . . 73
fast Fourier transform . . . . . . . . . . . . . . . . 78
frequency domain . . . . . . . . . . . . . . . . . . 79
sine (waves) . . . . . . . . . . . . . . . . . . . . . 73
tilting . . . . . . . . . . . . . . . . . . . . . . . . 79
time domain . . . . . . . . . . . . . . . . . . . . . 79
Fuzzy clustering . . . . . . . . . . . . . . . . . . . 142, 256

## G

Generalised linear model (GLM) . . . . . . . . . . . . . . 150
Genetic algorithms . . . . . . . . . . . . . . . . . . . . . . 63
    combination . . . . . . . . . . . . . . . . . . . . . . 64
    fitness . . . . . . . . . . . . . . . . . . . . . . . . . 65
    mutation . . . . . . . . . . . . . . . . . . . . . . . . 65
    selection . . . . . . . . . . . . . . . . . . . . . . . . 66

## I

Influence, influential . . . . . . . . . . . . . . . . . . 179, 187
    Cook's influence measure . . . . . . . . . . . . . . 188
Instrument standardisation . . . . . . . . . . . . . . . . . 207
    bias and slope correction . . . . . . . . . . . . . . . 217
    direct standardisation . . . . . . . . . . . . . . . . 209
    master . . . . . . . . . . . . . . . . . . . . . . . . . 208
    slave . . . . . . . . . . . . . . . . . . . . . . . . . 209
    piecewise direct standardisation . . . . . . . . . . . 211
    robust calibration . . . . . . . . . . . . . . . . . . 216
Intercept . . . . . . . . . . . . . . . . . . . . . . . . . 29, 299
Inverse calibration . . . . . . . . . . . . . . . . . . . . . 11

## J

Jack-knife . . . . . . . . . . . . . . . . . . . . . . . . . . 67

## K

Kubelka–Munck transformation . . . . . . . . . . . . 98, 106

## L

Latent variable . . . . . . . . . . . . . . . . . . . . . . . 39
    see also Principal component analysis
Least squares (LS) . . . . . . . . . . . . . . . . 11, 19, 56, 299
    weighted least squares (WLS) . . . . . . . . . . . . 315
Leverage . . . . . . . . . . . . . . . . . . . . . . . . . . 181
Linear model . . . . . . . . . . . . . . . . . . . . 56, 299, 311
    regression . . . . . . . . . . . . . . . . . . . . . . 298
Loadings . . . . . . . . . . . . . . . . . . . . . 29, 31, 34, 39
    see also Principal component analysis (PCA) and Partial least squares (PLS) regression

Loading weights . . . . . . . . . . . . . . . . . . . . . 33, 42, 43
    see also Partial least squares (PLS) regression
Locally weighted regression (LWR). . . . . . . . 45, 102, 127
    CARNAC . . . . . . . . . . . . . . . . . . . . . . . 133
    (Comparison analysis using restructured near infrared and constituent data)
    distance measures . . . . . . . . . . . . . . . . . . . . 131

## M

Mahalanobis distance . . . . . . . . . 131, 227, 228, 237, 241
    truncated . . . . . . . . . . . . . . . . . . . . . . . . 132
Matrix algebra, technical details . . . . . . . . . . . . . . 285
    determinant . . . . . . . . . . . . . . . . . . . . . . 228
    eigenvalue . . . . . . . . . . . . . . . . . . . . . . . 294
    eigenvector . . . . . . . . . . . . . . . . . . . . . . . 294
    full rank . . . . . . . . . . . . . . . . . . . . . . . . 292
    identity matrix . . . . . . . . . . . . . . . . . . . . . 291
    inner product . . . . . . . . . . . . . . . . . . . . . . 292
    inverse . . . . . . . . . . . . . . . . . . . . . . . . . 291
    linear transformation . . . . . . . . . . . . . . . . . . 291
    matrix . . . . . . . . . . . . . . . . . . . . . . . . . 288
    multiplication . . . . . . . . . . . . . . . . . . . . . 290
    orthogonal matrix . . . . . . . . . . . . . . . . . . . 293
    projection . . . . . . . . . . . . . . . . . . . . . . . 293
    singular value . . . . . . . . . . . . . . . . . . . . . 295
    singular value decomposition (SVD) . . . . . . . . . 295
    square matrix . . . . . . . . . . . . . . . . . . . . . . 294
    summation . . . . . . . . . . . . . . . . . . . . . . . 289
    transpose . . . . . . . . . . . . . . . . . . . . . . . . 286
    vectors . . . . . . . . . . . . . . . . . . . . . . . . . 285
Mean-centred . . . . . . . . . . . . . . . . . . . . . . 27, 295
Mean square error (MSE) . . . . . . . . . . . . . . . . . . 12
    of bootstrapping (MSEBT) . . . . . . . . . . . . . . 163
    of prediction (MSEP) . . . . . . . . . . . . . . . . . . 57
Monitoring calibration equations . . . . . . . . . . . . . . 201
    statistical process control (SPC) . . . . . . . . . . . . 202
    Shewhart chart . . . . . . . . . . . . . . . . . . . . . 203
Multicollinearity . . . . . . . . . . . . . . . . . . . . . 19, 234

exact multicollinearity . . . . . . . . . . . . . . . . . 19
measures of multicollinearity . . . . . . . . . . . . . 21
near-multicollinearity . . . . . . . . . . . . . . . . . . 19
Multiple linear regression (MLR) . . . . . . . . . 17, 150, 311
Multiplicative scatter/signal correction (MSC) . 15, 47, 99, 114
piecewise (PMSC) . . . . . . . . . . . . . . . . . . . 119
Multiplicative effect . . . . . . . . . . . . . . . . . 99, 106, 108
Multivariate . . . . . . . . . . . . . . . . . . . . . . . . . . . . 5
calibration . . . . . . . . . . . . . . . . . . . . . . . 5, 329
classification . . . . . . . . . . . . . . . . . . . . . 5, 332
linear model . . . . . . . . . . . . . . . . . . . . 19, 311
outlier . . . . . . . . . . . . . . . . . . . . . . . . . . . 178

# N

Near infrared spectroscopy (NIR) . . . . . . . . . . . . 1, 323
absorbance . . . . . . . . . . . . . . . . . . . . . 106, 324
combinations . . . . . . . . . . . . . . . . . . . . . . 325
hydrogen bonding . . . . . . . . . . . . . . . . . . 326
MIR and FIR . . . . . . . . . . . . . . . . . . . . . . 323
overtones . . . . . . . . . . . . . . . . . . . . . . . . 325
reflectance . . . . . . . . . . . . . . . . . . . . . 106, 324
transmission . . . . . . . . . . . . . . . . . . . . . . 324
Neural network (or nets) (NN) . . . . . . . . . . . . . 102, 146
back-propagation . . . . . . . . . . . . . . . . . . . 148
feed-forward networks . . . . . . . . . . . . . . 146, 148
hidden layer . . . . . . . . . . . . . . . . . . . . . . 146
input layer . . . . . . . . . . . . . . . . . . . . . . . 146
output layer . . . . . . . . . . . . . . . . . . . . . . 146
Sigmoid transformation function . . . . . . . . . . 147
Non-linearity . . . . . . . . . . . . . . . . . . . . . 6, 44, 93, 137
multivariate . . . . . . . . . . . . . . . . . . . . . . . 93
univariate . . . . . . . . . . . . . . . . . . . . . . . . 93
Non-linear calibration . . . . . . . . . . . . . . . . . . . . . 101
generalised linear model (GLM) . . . . . . . . . . . 150
locally weighted regression (LWR) . . . . . . . . . 127
neural networks (NN) . . . . . . . . . . . . . . . . 102
non-linear PLS regression . . . . . . . . . . . . . . 153
projection pursuit (PP) . . . . . . . . . . . . . . . . 152

splitting data into subgroups . . . . . . . . . . . . . . 140
transformations . . . . . . . . . . . . . . . . . . . 98, 99
using polynomial functions . . . . . . . . . . . . . 137
Non-selectivity . . . . . . . . . . . . . . . . . . . . . . 6, 14, 223

## O

Optimised scaling (OS) . . . . . . . . . . . . . . . . 99, 123
Orthogonal . . . . . . . . . . . . . . . . . . . . . . . . . 22, 40
    loadings . . . . . . . . . . . . . . . . . . . . . . . . 42
    scores . . . . . . . . . . . . . . . . . . . . . . . . . 42
    signal correction (OSC) . . . . . . . . . . . . . 99, 122
Orthogonal signal correction (OSC) . . . . . . . . . . 99, 122
Outlier . . . . . . . . . . . . . . . . . . . . . . . . . . . 6, 248
    detection . . . . . . . . . . . . . . . . . . . . . . 177
    leverage . . . . . . . . . . . . . . . . . . . . . . 181
    leverage outlier. . . . . . . . . . . . . . . . . . . 184
    multivariate . . . . . . . . . . . . . . . . . . . . 178
    univariate. . . . . . . . . . . . . . . . . . . . . . 178
    X-outlier . . . . . . . . . . . . . . . . . . . . . . 179
    X-residual . . . . . . . . . . . . . . . . . . . . . 184
    Y-outlier . . . . . . . . . . . . . . . . . . . . . . 179
    Y-residual. . . . . . . . . . . . . . . . . . 95, 101, 186
Overfitting . . . . . . . . . . . . . . . . . . . . . . . . . 24

## P

Partial least squares (PLS) regression . . . . . . . . . . . 33
    loadings . . . . . . . . . . . . . . . . . . . . . . . 34
    loadings weights. . . . . . . . . . . . . . . . 33, 42, 43
    scores . . . . . . . . . . . . . . . . . . . . . . . . 34
Precision. . . . . . . . . . . . . . . . . . . . . . . . . . 164
Prediction equation . . . . . . . . . . . . . . . . . . . 29, 32
    outlier . . . . . . . . . . . . . . . . . . . . . . . 178
    testing . . . . . . . . . . . . . . . . . . . . . . . 157
Principal component . . . . . . . . . . . . . . . . . . . . 30
    analysis. . . . . . . . . . . . . . . . . . . . . 30, 315
    cluster analysis. . . . . . . . . . . . . . . . . . . 249
    in classification . . . . . . . . . . . . . . . . . . 235
    loadings . . . . . . . . . . . . . . . . . . . . . 31, 39

regression . . . . . . . . . . . . . . . . . . . . . . 30
scores. . . . . . . . . . . . . . . . . . . . 31, 39, 237

# R

Reflectance. . . . . . . . . . . . . . . . . . . . 106, 324
Regression . . . . . . . . . . . . . . . . . . . . . 29, 32
    coefficients . . . . . . . . . . 29, 32, 68, 299, 308, 311
    continuum regression . . . . . . . . . . . . . . . . 37
    intercept . . . . . . . . . . . . . . . . . . . . . . 299
    model. . . . . . . . . . . . . . . . . . . . . . . . 299
    multiple linear regression (MLR) . . . . . . 17, 150, 311
    non-linear . . . . . . . . . . . . . . . . . . . . . . 97
    vector . . . . . . . . . . . . . . . . . . . . . . . . 29
Residual . . . . . . . . . . . . . . . . . 29, 301, 303, 312, 313
    matrix . . . . . . . . . . . . . . . . . . . . . . . . 39
    standard deviation . . . . . . . . . . . . . . . . . 306
    sum of squares (*RSS*). . . . . . . . . . . . . . 56, 301
    variance . . . . . . . . . . . . . . . . . . . . . . . 57
    X-residual . . . . . . . . . . . . . . . . . . . . . 184
    Y-residual. . . . . . . . . . . . . . . . . 95, 101, 186
Robust calibration . . . . . . . . . . . . . . . . . . . 216
    rugged calibration . . . . . . . . . . . . . . . . . 217
    robust wavelengths. . . . . . . . . . . . . . . . . 217
Root mean square error (*RMSE*) . . . . . . . . . . . . 155
    of bootstrapping (*RMSEBT*) . . . . . . . . . . . . 163
    of calibration (*RMSEC*) . . . . . . . . . . . . . . 156
    of cross-validation (*RMSECV*). . . . . . . . . . . 160
    of prediction (*RMSEP*). . . . . . . . . . . . . 59, 157

# S

Sample selection . . . . . . . . . . . . . . . . . 191, 212
Scatter . . . . . . . . . . . . . . . . . . . . . . . 15, 105
    additive component. . . . . . . . . . . . . . 106, 114
    correction . . . . . . . . . . . . . . . . 15, 47, 99, 105
    derivative. . . . . . . . . . . . . . . . . . . . . . 107
    effect (light scatter) . . . . . . . . . . . . . . . . 105
    multiplicative component. . . . . . . . . . . . 105, 114
    multiplicative correction (MSC) . . . . . . . . 15, 47, 99

optimised scaling (OS). . . . . . . . . . . . . . . . . 123
orthogonal signal correction (OSC) . . . . . . . . . . 122
pathlength correction method (PLC-MC) . . . . . . . 120
piecewise scatter correction (PMSC) . . . . . . . . . 119
standard normal variate (SNV) . . . . . . . . . . . 124
Scores . . . . . . . . . . . . . . . . . . . . 29, 31, 34, 39, 237
    see also Principal component analysis (PCA) and Partial least squares (PLS) regression
Semi-parametric regression . . . . . . . . . . . . . . . 98
Shrinking . . . . . . . . . . . . . . . . . . . . . . . 12, 140
Singular value decomposition (SVD). . . . . . . . . . . 295
Smoothing. . . . . . . . . . . . . . . . . . . . . . . . 110
Standard error of prediction (*SEP*) . . . . . . . . . . 164, 169
Standard normal variate (SNV) . . . . . . . . . . . . . 124
Standardisation . . . . . . . . . . . . . . . . . . . . . 29
    of x-variables . . . . . . . . . . . . . . . . . . . . 29
    of instruments . . . . . . . . . . . . . . . . . . . 207
    robust calibration . . . . . . . . . . . . . . . . . 216
Statistical process control (SPC) . . . . . . . . . . . . 202
    Shewhart chart . . . . . . . . . . . . . . . . . . . 203
Stein estimation . . . . . . . . . . . . . . . . . . . . 140
Stepwise regression. . . . . . . . . . . . . . . . . . . . 60
Sum of squares (SS) . . . . . . . . . . . . . . . . . . 31, 32

# T

t-test . . . . . . . . . . . . . . . . . . . . . . . . . 33, 314
Test set . . . . . . . . . . . . . . . . . . . . . . . . . 59
    see also Validation
Testing differences in prediction error. . . . . . . . . 167, 169
Transformations . . . . . . . . . . . . . . . . . . . . . 97
    Box–Cox. . . . . . . . . . . . . . . . . . . . . . . 100
    Kubelka–Munck. . . . . . . . . . . . . . . . . . . . 98
    spectroscopic . . . . . . . . . . . . . . . . . . . . 98
    statistical . . . . . . . . . . . . . . . . . . . . . . 99

# U

Underfitting. . . . . . . . . . . . . . . . . . . . . . . 24
Univariate calibration. . . . . . . . . . . . . . . . . . 11

## V

Validation . . . . . . . . . . . . . . . . . . . . 59, 155
    Akaike information criterion . . . . . . . . . . . . . 58
    bias . . . . . . . . . . . . . . . . . . . . . . . . . . . 164
    bootstrapping . . . . . . . . . . . . . . . . . . . . . 162
    cross-validation . . . . . . . . . . . . . . . . 59, 160
    of classification rules . . . . . . . . . . . . . . . . 247
    prediction testing . . . . . . . . . . . . . . . . . . . 157
    root mean square error (*RMSE*) . . . . . . . . . . . 155
Standard error of prediction . . . . . . . . . . . . . . . . 164
    test set . . . . . . . . . . . . . . . . . . . . . . . . . 59
    (see also root mean square error)
Variance . . . . . . . . . . . . . . . . . . . . . . . . . . 297
Variance inflation factor (VIF) . . . . . . . . . . . . . 21, 35
Variable selection . . . . . . . . . . . . . . . . . . . . . 55
    backward elimination . . . . . . . . . . . . . . . . . 60
    best subset selection . . . . . . . . . . . . . . . . . . 60
    forward selection . . . . . . . . . . . . . . . . . . . 60
    genetic algorithms . . . . . . . . . . . . . . . . . . . 63
    jack-knife . . . . . . . . . . . . . . . . . . . . . . . 67
    stochastic search . . . . . . . . . . . . . . . . . . . . 62

## W

Wavelets . . . . . . . . . . . . . . . . . . . . . . . 71, 84
Weight function . . . . . . . . . . . . . . . . . . . . . 133